Integral Equations
Via Imbedding Methods

APPLIED MATHEMATICS AND COMPUTATION

A Series of Graduate Textbooks, Monographs, Reference Works
Series Editor: ROBERT KALABA, University of Southern California

Integral Equations
Via
Imbedding Methods

HARRIET H. KAGIWADA
The Rand Corporation
Santa Monica, California

ROBERT KALABA
University of Southern California
Los Angeles, California

 1974

Addison-Wesley Publishing Company
Advanced Book Program
Reading, Massachusetts

London · Amsterdam · Don Mills, Ontario · Sydney · Tokyo

CODEN: APMCC

Library of Congress Cataloging in Publication Data

Kagiwada, Harriet H   1937–
    Integral equations via imbedding methods.

    (Applied mathematics and computation, no. 6)
    Includes bibliographical references.
    1. Integral equations. 2. Initial value problems.
3. Cauchy problem. 4. Invariant imbedding.
I. Kalaba, Robert E., joint author. II. Title.
QA431.K28   515'.45   74–13228
ISBN 0–201–04106–5
ISBN 0–201–04107–3 (pbk.)

Reproduced by Addison-Wesley Publishing Company, Inc., Advanced Book Program, Reading, Massachusetts, from camera-ready copy prepared by the authors.

American Mathematical Society (MOS) Subject Classification Scheme (1970): 47A55, 45B05, 65L10, 65R05

Manufactured in the United States of America

CONTENTS

SERIES EDITOR'S FOREWORD

Execution times of modern digital computers are measured in nano-seconds. They can solve hundreds of simultaneous ordinary differential equations with speed and accuracy. But what does this immense capability imply with regard to solving the scientific, engineering, economic, and social problems confronting mankind? Clearly, much effort has to be expended in finding answers to that question.

In some fields, it is not yet possible to write mathematical equations which accurately describe processes of interest. Here, the computer may be used simply to simulate a process and, perhaps, to observe the efficacy of different control processes. In others, a mathematical description may be available, but the equations are frequently difficult to solve numerically. In such cases, the difficulties may be faced squarely and possibly overcome; alternatively, formulations may be sought which are more compatible with the inherent capabilities of computers. Mathematics itself nourishes and is nourished by such developments.

Each order of magnitude increase in speed and memory size of computers requires a reexamination of computational techniques and an assessment of the new problems which may be brought within the realm of solution. Volumes in this series will provide indications of current thinking regarding problem formulations, mathematical analysis, and computational treatment.

ROBERT KALABA

PREFACE

Traditionally, there has been a sharp dichotomy between the study of initial value problems, on the one hand, and boundary value problems and integral equations, on the other. This book may help tear down these distinctions by showing how to transform integral equations into initial value problems or Cauchy systems. The method applies to linear and nonlinear integral equations, singular integral equations, eigenvalue problems for integral operators and to bifurcation problems.

There is, of course, an analytic advantage in having new equations with which to study integral equations. But we also have new computational methods eminently suited to the ability of modern analog and digital computers to resolve initial value problems. This book is merely suggestive, many matters of rigor and computational optimization being left open.

The general importance of integral equations in applied mathematics is indicated in the bibliography of B. Noble, *A Bibliography on Methods for Solving Integral Equations*, Mathematics Research Center, University of Wisconsin, Madison, 1971. While this book is largely self-contained, it would be convenient to have on hand a copy of Courant-Hilbert, *Methods of Mathematical Physics*, Vol. 1, Interscience, 1957, and of J. Cochran, *Analysis of Linear Integral Equations*, McGraw-Hill, 1972. There is an extensive list of references to imbedding methods in applied mathematics in M. Scott's *A Bibliography on Invariant Imbedding and Related Topics*, Sandia Laboratories Report SC-B-71-0886, Albuquerque, New Mexico.

Let us consider the linear Fredholm integral equation,

$$u(t) = g(t) + \int_0^x k(t,y) \, u(y) \, dy \, ,$$

$$0 \leq t \leq x.$$

We imbed this single integral equation within a class of such equations by considering the interval length x to be a variable, $0 \leq t \leq x$, so that

$$u = u(t) = u(t,x) \, ,$$

$$0 \leq t \leq x \leq X,$$

and we assume that there is a unique solution when X is sufficiently small. We shall derive a complete Cauchy system for the function u and for certain auxiliary functions in which the interval length x plays the role of the time-like variable. For nonlinear integral equations, similar reductions can be made. Thus integral equations can be conceptually related to the theory of semi-groups (E. Hille and R. Phillips, *Functional Analysis and Semi-Groups,* American Mathematical Society, Providence, Rhode Island, 1957). While matters of rigor remain, a strong start on this front has been made by M. Golberg.

The computational implementation of the imbedding method is successful for two prime reasons. The numerical integration of systems of ordinary differential equations with complete sets of initial conditions is readily done on electronic computers using schemes of varying accuracy and efficiency (L. Collatz, *The Numerical Treatment of Differential Equations,* Springer-Verlag, Berlin, 1960) provided the

equations are stable. And indeed, basic imbedding equations
do turn out to be stable.

In 1943 V. Ambarzumian published a famous paper on
multiple scattering. He considered parallel rays of light
incident on a homogeneous layer bounded by two parallel planes
and within which the light is absorbed and re-emitted isotropically. He showed that it is possible to determine the
diffusely reflected radiation without determining the entire
radiation field within the slab. He did this by observing
that the reflected radiation is unchanged if we add a layer
of a certain thickness to the top and delete an identical
layer from the bottom of the slab. As is discussed in Chapter 9 of this book, the underlying functional equation is a
Fredholm integral equation, and Ambarzumian was showing how
a functional on its solution varies as a function of the interval length. V. V. Sobolev, who has contributed so much
to the theory, has an excellent discussion in his book, *A
Treatise on Radiative Transfer*, D. Van Nostrand Co., Princeton, New Jersey, 1963. Sobolev, himself, has greatly in-
fluenced our work by pointing out the importance of the function Φ (see Chapters 4 and 9).

We began our imbedding studies following the physical
type of approach used by Ambarzumian, Sobolev, R. Bellman,
S. Ueno, and G. M. Wing. We derived the initial value problem for the integral equation of radiative transfer, first
physically then analytically. We succeeded in extending the
method to general displacement kernels having the form,

$$k(t,y) = k(|t-y|)$$

$$= \int_0^x e^{-|t-y|/z} \, w(z) \, dz \, ,$$

and to other classes of kernels. In all of this, the length
of the interval was the independent variable.

The richest result in this part of the theory concerns
the functions b and h. For integral equations with displacement kernels, certain functionals of the solution may
well be the quantities of interest. The functions b and h,
while of interest in themselves, do provide simple algebraic
means for determining these functionals. The functions
b and h are calculated as solutions of initial value problems, while they also satisfy a system of singular integral
equations. The analytical discussion is given in Chapter 4,
and the physical basis is shown in Chapter 9.

Another type of imbedding is the imbedding of the
original integral equation in a class of integral equations
having a parameter, say λ:

$$u(t,\lambda) = g(t) + \int_0^1 k(t,y,\lambda,u(y,\lambda))\ dy\ ,$$
$$0 \leq t \leq 1,\quad 0 \leq \lambda \leq \Lambda,$$

where both u and λ may appear nonlinearly under the integral. It also leads to an effective procedure for the
determination of eigenvalues and critical lengths, as
Chapter 8 reveals.

A bonus not to be overlooked in casting integral equations into a Cauchy mold is the accessibility of solving inverse problems by a variety of system identification techniques. When there are unknown parameters in the integral
equation to be determined by some measurements on the solution, then imbedding together with a method such as quasilinearization can produce good numerical results. This is

shown by the example from radiative transfer in Chapter 9.

In a course or seminar on imbedding methods in integral equations, one would probably begin with Chapters 1 and 2. These are concerned with degenerate kernels,

$$k(t,y) = \sum_{m=1}^{M} g_m(t) \, h_m(y) \, ,$$

$$0 \leq t, y \leq x,$$

and semidegenerage kernels,

$$k(t,y) = \begin{cases} \displaystyle\sum_{i=1}^{M} \alpha_i(t) \, \beta_i(y) \, , & 0 \leq y \leq t, \\[3em] \displaystyle\sum_{j=1}^{N} \gamma_j(t) \, \delta_j(y) \, , & t < y \leq x. \end{cases}$$

These chapters provide a number of illustrative examples and a listing of a working computer program. Then, one might study parts of Chapters 4 and 9, to be followed by Chapter 7 which is a general discussion of interval imbedding for nonlinear integral equations. Chapter 8 treats nonlinear integral equations and various special cases by parameter imbedding. Finally, Chapter 10 is an imbedding approach to dual integral equations for the function $A(\lambda)$,

$$\int_0^\infty A(\lambda) \, J_o(\lambda r) \, d\lambda = f(r) \, , \qquad r < a,$$

$$\int_0^\infty \frac{\lambda \, A(\lambda)}{1 - g(\lambda)} \, J_o(\lambda r) \, d\lambda = 0 \, , \qquad r > a,$$

which is of importance in the study of mixed boundary value
problems of potential theory. Some exercises are provided at
the ends of the chapters.

Names of some of those who have worked on imbedding of
integral equations, whether analytically or computationally,
will be found in the pages of this book. We take the oppor-
tunity here to express our appreciation to M. R. Scott for
his careful reading of this manuscript, and to L. Anderson
and S. Whitaker for the typing.

In this age of computers we are concerned not only
with analytical elegance but with effective numerical
treatment of realistic nonlinear models of physical and social
phenomena. Imbedding methods now offer an attractive approach
in both areas, and may grow in significance in the coming de-
cades. It is with this expectation that we present this work

Harriet H. Kagiwada
Robert E. Kalaba

CHAPTER 1

DEGENERATE KERNELS

1.1 INTRODUCTION

The mathematical treatment of many problems in physics and engineering concerns the solution of the Fredholm integral equation,

$$u(t) = f(t) + \int_0^a k(t,y) \, u(y) \, dy \, , \qquad (1.1)$$

$$0 \leq t \leq a,$$

where $k(t,y)$ is the kernel, $f(t)$ is the forcing function, and $u(t)$ is the solution on the fixed interval, $0 \leq t \leq a$. In this chapter we shall consider kernels which are degenerate, that is, kernels which have the form

$$k(t,y) = \sum_{m=1}^{M} g_m(t) \, h_m(y) \, , \qquad (1.2)$$

$$0 \leq t, y \leq a.$$

It is assumed that the functions $\{g_m\}$ and $\{h_m\}$ are sufficiently smooth functions of their arguments.

1

The standard procedure in solving Eq. (1.1) is to convert it into an equivalent matrix equation. The solution is then computed by evaluating a number of integrals and performing a matrix inversion. This last step has computational difficulties.

In this first chapter, we show how to convert the integral equation into an equivalent initial value problem, which can be solved effectively by an analog or digital computer. Before doing so, however, we take up a few examples for which the solutions can explicitly be found. We then derive an initial value problem for one of these cases by observing that the solution depends upon the interval length. We can verify that the analytical solution does indeed satisfy this initial value problem.

We then derive the initial value problem for a general degenerate kernel. This leads to a numerical method which is extremely accurate and effective, as several computational experiments indicate. It is also useful in the calculation of eigenvalues or critical lengths. Next, it is verified that the solution of the Cauchy system does satisfy the integral equation. Then a representation property of the resolvent kernel is derived.

The initial value problem for the case of the degenerate kernel is taken up first in this book because a differential equation for the solution itself is not required. Rather, the solution is obtained with the aid of a simple representation formula. This is not possible for the kernel which we consider in the following chapter.

1.2   SOME EXAMPLES

Consider the integral equation

$$u(t) = 1 + \int_0^x e^{-(t+y)} u(y) \, dy \, , \tag{1.3}$$

$$0 \leq t \leq x$$

It may be rewritten as

$$u(t) = 1 + e^{-t} \int_0^x e^{-y} u(y) \, dy \, , \tag{1.4}$$

$$0 \leq t \leq x.$$

Calling

$$c(x) = \int_0^x e^{-y} u(y) \, dy \, , \tag{1.5}$$

it takes the form

$$u(t) = 1 + c(x) e^{-t} \, , \tag{1.6}$$

$$0 \leq t \leq x.$$

To complete the solution, we have but to determine the function c.   To do this we write

$$c(x) = \int_0^x e^{-y} \left[1 + e^{-y} c(x) \right] dy \, . \tag{1.7}$$

It follows that

$$c(x) = 1 - e^{-x} + c(x) \left[1 - e^{-2x} \right]/2 \, . \tag{1.8}$$

Solving the last equation for c(x) we find that

$$c(x) = \frac{2-2e^{-x}}{1+e^{-2x}} \, ,$$

(1.9)

$$0 \leq x < \infty.$$

The solution of the integral equation is

$$u(t) = 1 + \frac{2-2e^{-x}}{1+e^{-2x}} e^{-t} \, ,$$

(1.10)

$$0 \leq t \leq x < \infty.$$

It is really no more difficult to solve the integral equation with a general forcing term,

$$u(t) = f(t) + \int_0^x e^{-(t+y)} u(y) \, dy \, ,$$

(1.11)

$$0 \leq t \leq x.$$

We write

$$u(t) = f(t) + e^{-t} \int_0^x e^{-y} u(y) \, dy$$

(1.12)

$$= f(t) + e^{-t} c(x) \, ,$$

where

$$c(x) = \int_0^x e^{-y} u(y) \, dy \, .$$

(1.13)

To find $c(x)$ we proceed as before and obtain

$$c(x) = \int_0^x e^{-y} \left[f(y) + e^{-y} c(x) \right] dy$$

$$= \int_0^x e^{-y} f(y) \, dy + c(x) \int_0^x e^{-2y} \, dy$$

(1.14)

$$= \int_0^x e^{-y} f(y) \, dy + c(x) \left[\frac{1}{2} + \frac{e^{-2x}}{2} \right] \, .$$

The result is that

$$c(x) = \frac{2 \int_0^x e^{-y} f(y) dy}{1+e^{-2x}} , \qquad (1.15)$$

$$0 \le x < \infty.$$

The solution of the integral equation is

$$u(t) = f(t) + e^{-t} \frac{2 \int_0^x e^{-y} f(y) dy}{1+e^{-2x}} , \qquad (1.16)$$

$$0 \le t \le x < \infty.$$

Notice that the solution may be rewritten in the form

$$u(t) = f(t) + \int_0^x \left[\frac{2e^{-(t+y)}}{1+e^{-2x}} \right] f(y) dy . \qquad (1.17)$$

Introducing the resolvent kernel K,

$$K(t,y) = \frac{2e^{-(t+y)}}{1+e^{-2x}} , \qquad (1.18)$$

we see that the solution may be written in the form

$$u(t) = f(t) + \int_0^x K(t,y) f(y) dy , \qquad (1.19)$$

which shows the explicit linear dependence of the solution u upon the forcing term f. Observe also that the solution is valid for all non-negative interval lenths.

For the slightly different integral equation

$$u(t) = 1 + \int_0^x e^{(t+y)} u(y) \, dy \, ,$$

(1.20)

$$0 \le t \le x,$$

the behavior of the solution is quite different. We find
that

$$u(t) = 1 + e^t \int_0^x e^y u(y) \, dy = 1 + c(x) \, e^t \, ,$$

(1.21)

where

$$c(x) = \int_0^x e^y u(y) \, dy \, .$$

(1.22)

Solving for $c(x)$ we find

$$c(x) = \int_0^x e^y \left[1 + c(x) \, e^y \right] dy$$
$$= e^x - 1 + c(x) \left[e^{2x} - 1 \right] / 2 \, ,$$

or

$$c(x) = \frac{2e^x - 2}{3 - e^{2x}} \, .$$

(1.23)

The last equation holds provided that

$$3 - e^{2x} \neq 0 \, .$$

(1.24)

The positive root, x_1, of the equation

$$3 - e^{2x} = 0$$

(1.25)

is

$$x_1 = (1/2) \ln 3 = 0.54930615 . \qquad (1.26)$$

It follows that the solution of the integral equation is

$$u(t) = 1 + \frac{2e^x - 2}{3 - e^{2x}} e^t , \qquad (1.27)$$

for

$$0 \le t \le x \ne x_1 .$$

Next consider the integral equation with the same kernel but a general forcing term,

$$u(t) = f(t) + \int_0^x e^{(t+y)} u(y) \, dy , \qquad (1.28)$$

$$0 \le t \le x.$$

To solve it we write

$$u(t) = f(t) + e^t \int_0^x e^y u(y) \, dy$$
$$= f(t) + e^t c(x) , \qquad (1.29)$$

where

$$c(x) = \int_0^x e^y u(y) \, dy . \qquad (1.30)$$

Again substitution from the previous equation yields

$$c(x) = \int_0^x e^y \left[f(y) + e^y c(x) \right] dy$$

$$= \int_0^x e^y f(y) dy + c(x) \int e^{2y} dy .$$

(1.31)

It follows that

$$c(x) = \frac{2 \int_0^x e^y f(y) dy}{3-e^{2x}} ,$$

(1.32)

provided that the denominator is not zero,

$$x \neq x_1 .$$

(1.33)

The solution of the integral equation is

$$u(t) = f(t) + e^t \frac{2 \int_0^x e^y f(y) dy}{3-e^{2x}} ,$$

(1.34)

for

$$0 \leq t \leq x \neq x_1 .$$

(1.35)

We shall refer to an interval length x for which the integral equation does not have a unique solution as a critical length. Observe that for $x \neq x_1$, we may write the solution in the form

$$u(t) = f(t) + \int_0^x \frac{2e^{t+y}}{3-e^{2x}} f(y) dy ,$$

(1.36)

$$0 \leq t \leq x \neq x_1 .$$

In this case the resolvent kernel K is

$$K(t,y) = \frac{2e^{t+y}}{3-e^{2x}} ,$$

$$x \neq x_1 . \tag{1.37}$$

1.3   INITIAL VALUE TREATMENT

Let us now see how we may reduce the solution of one
of the previous examples to the solving of ordinary differential equations with known initial conditions. We observe
that the solution of the integral equation

$$u(t) = f(t) + \int_0^x e^{-(t+y)} u(y) \, dy , \tag{1.38}$$

$$0 \leq t \leq x < \infty,$$

actually depends upon the interval length x, as well as upon
the variable t. Let us then write

$$u = u(t) = u(t,x) , \tag{1.39}$$

$$0 \leq t \leq x < \infty,$$

and

$$u(t,x) = f(t) + \int_0^x e^{-(t+y)} u(y,x) \, dy . \tag{1.40}$$

We differentiate both sides of the last equation with respect
to x to obtain

$$u_x(t,x) = e^{-(t+x)} u(x,x) + \int_0^x e^{-(t+y)} u_x(y,x) \, dy \, , \tag{1.41}$$

where we have employed the notation

$$u_x(t,x) = \partial u/\partial x \, . \tag{1.42}$$

We also have used the rule for differentiating an integral with respect to a variable which appears in the integrand and in the upper limit,

$$\frac{d}{dx} \int_0^x F(y,x) \, dy = F(x,x) + \int_0^x \frac{\partial F}{\partial x} \, dy \, . \tag{1.43}$$

Next we introduce the auxiliary function Φ as the solution of the integral equation

$$\Phi(t,x) = e^{-(t+x)} + \int_0^x e^{-(t+y)} \Phi(y,x) \, dy \, , \tag{1.44}$$
$$0 \leq t \leq x \, .$$

We regard the earlier equation involving u_x as an integral equation for u_x having as its forcing term

$$e^{-(t+x)} \, .$$

By comparing the forcing terms of the integral equations for the functions u_x and Φ and noting that $u(x,x)$ does not depend upon t , we see that

$$u_x(t,x) = \Phi(t,x) \, u(x,x) \, ,$$

(1.45)

$$0 \leq t \leq x.$$

Introduce the new auxiliary function J to be the solution of the integral equation

$$J(t,x) = e^{-t} + \int_0^x e^{-(t+y)} \, J(y,x) \, dy \, ,$$

(1.46)

$$0 \leq t \leq x.$$

Again by comparing the integral equations for the functions J and Φ, we find that they are connected by the relation

$$\Phi(t,x) = J(t,x) \, e^{-x} \, ,$$

(1.47)

$$0 \leq t \leq x.$$

Differentiation of both sides of the integral equation for the function J above shows that J_x satisfies the integral equation

$$J_x(t,x) = e^{-(t+x)} \, J(x,x) + \int_0^x e^{-(t+y)} \, J_x(y,x) \, dy \, ,$$

(1.48)

$$0 \leq t \leq x.$$

Comparing the integral equations for the functions J_x and Φ, we find that

$$J_x(t,x) = \Phi(t,x) \, J(x,x) \, ,$$

(1.49)

$$0 \leq t \leq x.$$

As in our earlier discussion, let us write

$$u(t,x) = f(t) + e^{-t} c(x) , \qquad (1.50)$$

where

$$c(x) = \int_0^x e^{-y} u(y,x) \, dy , \qquad (1.51)$$
$$0 \leq x.$$

By differentiating both sides of this last equation with respect to x we find that

$$c'(x) = e^{-x} u(x,x) + \int_0^x e^{-x} u_x(y,x) \, dy , \qquad (1.52)$$

where

$$c'(x) = dc/dx . \qquad (1.53)$$

Using our earlier relation for u_x we see that

$$c'(x) = e^{-x} u(x,x) + \int_0^x e^{-y} \Phi(y,x) \, u(x,x) \, dy$$
$$= u(x,x) \left[e^{-x} + \int_0^x e^{-y} \Phi(y,x) \, dy \right] . \qquad (1.54)$$

From the integral equation for the function u itself for $t = x$, we see that

$$u(x,x) = f(x) + e^{-x} c(x) . \qquad (1.55)$$

To evaluate the integral in the earlier equation we use the

relation between the functions Φ and J ,

$$\int_0^x e^{-y} \Phi(y,x) \, dy = \int_0^x e^{-y} J(y,x) \, e^{-x} \, dy = e^{-x} r(x) \, ,$$

$$(1.56)$$

where we have introduced the final auxiliary function r by means of the definition

$$r(x) = \int_0^x e^{-y} J(y,x) \, dy \, , \qquad\qquad 0 \leq x \, . \qquad (1.57)$$

It follows that the differential equation for the function c is

$$c'(x) = \left[f(x) + e^{-x} c(x) \right] \left[e^{-x} + e^{-x} r(x) \right] \, , \qquad (1.58)$$

$$0 \leq x \, .$$

Furthermore, from the definition of the function c we see that the initial condition at $x = 0$ is

$$c(0) = 0 \, . \qquad\qquad\qquad\qquad (1.59)$$

The differential equation for the function c contains the function r . To round out the system of differential equations, we shall now obtain a differential equation for the function r and an initial condition at $x = 0$. By differentiating both sides of the equation defining the function r in terms of the function J with respect to x , we find the relation

$$r'(x) = e^{-x} J(x,x) + \int_0^x e^{-y} J_x(y,x) \, dy \, . \tag{1.60}$$

Earlier we showed that

$$J_x(t,x) = \Phi(t,x) \, J(x,x) \, , \tag{1.61}$$

so that

$$r'(x) = J(x,x) \left[e^{-x} + \int_0^x e^{-y} \, \Phi(y,x) \, dy \right] . \tag{1.62}$$

The integral in the above equation was evaluated previously, and the integral equation that defines the function J for t = x shows that

$$J(x,x) = e^{-x} + \int_0^x e^{-x+y} J(y,x) \, dy = e^{-x} + e^{-x} r(x) \, . \tag{1.63}$$

It follows that

$$r'(x) = \left[e^{-x} + e^{-x} r(x) \right] \left[e^{-x} + e^{-x} r(x) \right] \, , \tag{1.64}$$

which is the desired differential equation for the function r. From its definition, we see that the initial condition at x = 0 is

$$r(0) = 0 \, .$$

Let us now summarize the Cauchy system for the functions c and r. The differential equations are

$$c'(x) = [f(x) + e^{-x} c(x)] [e^{-x} + e^{-x} r(x)] , \qquad (1.65)$$

$$r'(x) = [e^{-x} + e^{-x} r(x)] [e^{-x} + e^{-x} r(x)] , \qquad (1.66)$$

valid for $x \geq 0$, and the initial conditions at $x = 0$ are

$$c(0) = 0 , \qquad\qquad\qquad\qquad\qquad (1.67)$$

$$r(0) = 0 . \qquad\qquad\qquad\qquad\qquad (1.68)$$

The function u is given by

$$u(t,x) = f(t) + e^{-t} c(x) , \qquad\qquad (1.69)$$

$$0 \leq t \leq x.$$

As a check we may solve for the functions r and c explicitly. We see that

$$\frac{dr}{dx} = e^{-2x} [1 + r^2] , \qquad r(0)=0 , \qquad\qquad (1.70)$$

which implies that

$$r(x) = \frac{1-e^{-2x}}{1+e^{-2x}} , \qquad\qquad\qquad (1.71)$$

$$0 \leq x.$$

This is seen by the separation of variables:

$$\int_0^x \frac{dr}{(1+r)^2} = \int_0^x e^{-2x} \, dx \; ,$$

$$-(1+r)^{-1} = \frac{-1}{2} e^{-2x} + k$$

$$= -\frac{1}{2} e^{-2x} - \frac{1}{2} \qquad\qquad (1.72)$$

$$\frac{1}{1+r} = \frac{1+e^{-2x}}{2} \; ,$$

$$1+r = \frac{2}{1+e^{-2x}} \; ,$$

from which the above formula for r follows.

To obtain the function c , we observe that it satisfies a first-order linear ordinary differential equation of
the form

$$\frac{dc}{dx} = p(x) \; c + q(x) \; , \qquad\qquad (1.73)$$

with the initial condition

$$c(0) = 0 \; . \qquad\qquad (1.74)$$

The solution is

$$c(x) = \int_0^x q(y) \; e^{\int_y^x p(x) \, ds} \; dy \; . \qquad\qquad (1.75)$$

By substituting

$$p(s) = \frac{2e^{-2s}}{1+e^{-2s}} \tag{1.76}$$

and

$$q(y) = \frac{2e^{-y}f(y)}{1+e^{-2y}} \, , \tag{1.77}$$

we find that

$$c(x) = \int_0^x \frac{2e^{-y}f(y)}{1+e^{-2x}} \, dy \, , \tag{1.78}$$

which is in complete agreement with what we found in the previous section for this function.

We also see that

$$
\begin{aligned}
J(t,x) &= e^{-t} + e^{-t} \int_0^x e^{-y} \, J(y,x) \, dy \\
&= e^{-t} + e^{-t} \, r(x) \\
&= e^{-t} \left[1 + r(x) \right] \\
&= \frac{2e^{-t}}{1+e^{-2x}} \, ,
\end{aligned} \tag{1.79}
$$

$$\Phi(t,x) = \frac{2e^{-(t+x)}}{1+e^{-2x}} \, ,$$

and

$$u(t,x) = f(t) + e^{-t} \frac{2\int_0^x e^{-y}f(y)\,dy}{1+e^{-2x}}, \tag{1.81}$$

as was found in the previous section.

1.4   GENERAL DERIVATION OF CAUCHY SYSTEM

Consider the integral equation

$$u(t) = f(t) + \int_0^a k(t,y)\,u(y)\,dy, \tag{1.82}$$

$$0 \le t \le a,$$

where the kernel k has the form

$$k(t,y) = \sum_{m=1}^{M} g_m(t)\,h_m(y). \tag{1.83}$$

Then Eq. (1.82) may be written

$$u(t) = f(t) + \sum_{m=1}^{M} g_m(t) \int_0^a h_m(y)\,u(y)\,dy. \tag{1.84}$$

Noting that the integral is a function only of the upper limit, a, for a given i , we define $c_i = c_i(a)$,

$$c_i(a) = \int_0^a h_i(y)\,u(y)\,dy, \tag{1.85}$$

$$i=1,2,\ldots,M.$$

If the coefficients c_i are known, then the solution of the integral equation is given by

$$u(t) = f(t) + \sum_{m=1}^{M} g_m(t) \, c_m \, . \qquad (1.86)$$

Substitution of Eq. (1.86) into Eq. (1.85) gives a system of linear algebraic equations for the unknowns c_i ,

$$c_i = \int_0^a h_i(y) \, f(y) \, dy + \sum_{m=1}^{M} c_m \int_0^a h_i(y) \, g_m(y) \, dy \, , \qquad (1.87)$$
$$i=1,2,\ldots,M \, .$$

The system (1.87) may be used to determine $c_1, c_2, \ldots, c_M$.

Since it is desirable to have more than one method of solution, we propose an alternate procedure for the determination of $u(t)$. We use an imbedding approach in which the solution at a fixed value of t is studied as the interval length is varied. Thus, we write the integral equation as

$$u(t,x) = f(t) + \int_0^x k(t,y) \, u(y,x) \, dy \, , \qquad (1.88)$$
$$0 \le t \le x \le x_1 \, ,$$

and

$$u(t,x) = f(t) + \sum_{m=1}^{M} g_m(t) \, c_m(x) \, , \qquad (1.89)$$

$$c_i(x) = \int_0^x h_i(y)\, u(y,x)\, dy \; ,$$

(1.90)

$$i = 1, 2, \ldots, M \; .$$

It is assumed that unity is not an eigenvalue of Eq. (1.88) for x between 0 and x_1 . See Section 1.5.

A Cauchy problem is derived for the function $c_i(x)$ in the following section. Then a brief summary of the initial value method is followed by numerical examples and a sufficiency proof for the Cauchy system.

We begin by differentiating Eq. (1.88) with respect to x , and we obtain

$$u_x(t,x) = k(t,x)\, u(x,x) + \int_0^x k(t,y)\, u_x(y,x)\, dy \; .$$

(1.91)

We regard this as an integral equation for $u_x(t,x)$. If $\Phi(t,x)$ is introduced as the solution of the integral equation

$$\Phi(t,x) = k(t,x) + \int_0^x k(t,y)\, \Phi(y,x)\, dy \; ,$$

(1.92)

then by the superposition principle we may express $u_x(t,x)$ as

$$u_x(t,x) = u(x,x)\, \Phi(t,x) \; .$$

(1.93)

The function $\Phi(t,x)$ may itself be expressed as

$$\Phi(t,x) = \sum_{m=1}^{M} h_m(x)\, J_m(t,x) \; ,$$

(1.94)

where $J_i(t,x)$ satisfies the equation

$$J_i(t,x) = g_i(t) + \int_0^x k(t,y) \, J_i(y,x) \, dy \; , \qquad (1.95)$$

$$i=1,2,\ldots,M \; .$$

Now differentiate Eq. (1.90) to obtain

$$c_i'(x) = h_i(x) \, u(x,x) + \int_0^x h_i(y) \, u_x(y,x) \, dy \; ,$$

and use Eq. (1.93). Then

$$c_i'(x) = u(x,x) \left\{ h_i(x) + \int_0^x h_i(y) \, \Phi(y,x) \, dy \right\} \; .$$

Use of Eq. (1.94) in the integral term of this equation gives

$$\int_0^x h_i(y) \, \Phi(y,x) \, dy = \sum_{m=1}^M h_m(x) \int_0^x h_i(y) \, J_m(y,x) \, dy \; ,$$

$$(1.96)$$

or

$$\int_0^x h_i(y) \, \Phi(y,x) \, dy = \sum_{m=1}^M h_m(x) \, r_{im}(x) \; , \qquad (1.97)$$

where $r_{ij}(x)$ is defined to be

$$r_{ij}(x) = \int_0^x h_i(y) \, J_j(y,x) \, dy \; , \qquad (1.98)$$

$$i,j=1,2,\ldots,M \; .$$

Then

$$c_i'(x) = u(x,x) \left\{ h_i(x) + \sum_{m=1}^{M} h_m(x) \, r_{im}(x) \right\} . \qquad (1.99)$$

For $u(x,x)$ we have from Eq. (1.89)

$$u(x,x) = f(x) + \sum_{m=1}^{M} g_m(x) \, c_m(x) . \qquad (1.100)$$

Equation (1.99) becomes

$$c_i'(x) = \left\{ f(x) + \sum_{m=1}^{M} g_m(x) \, c_m(x) \right\}$$

$$\cdot \left\{ h_i(x) + \sum_{m=1}^{M} h_m(x) \, r_{im}(x) \right\} . \qquad (1.101)$$

From the definition it is seen that

$$c_i(0) = 0 . \qquad (1.102)$$

Equations (1.101) and (1.102) form a system of differential equations and initial conditions for $c_i(x)$, for $i = 1, 2, \ldots, M$, and $0 \le x \le x_1$; i.e., they define the Cauchy problem for the $c_i(x)$. Next we show how to determine the function $r_{im}(x)$ which appears in Eq. (1.101).

Differentiation of Eq. (1.98) for $r_{ij}(x)$ gives

$$r_{ij}'(x) = h_i(x) \, J_j(x,x) + \int_0^x h_i(y) \, J_j'(y,x) \, dy , \qquad (1.103)$$

where the primes denote differentiation with respect to x. But it follows from differentiation of Eq. (1.95) that

$$J_j'(t,x) = J_j(x,x) \; \Phi(t,x) .$$
(1.104)

Use of Eqs. (1.104) and (1.97) yields

$$r_{ij}'(x) = J_j(x,x) \left\{ h_i(x) + \sum_{m=1}^{M} h_m(x) \; r_{im}(x) \right\} .$$
(1.105)

Observe that $J_j(t,x)$ is expressible in terms of $r_{mj}(x)$,

$$J_j(t,x) = g_j(t) + \sum_{m} g_m(t) \int_0^x h_m(y) \; J_j(y,x) \; dy$$

$$= g_j(t) + \sum_{m} g_m(t) \; r_{mj}(x) .$$
(1.106)

Equation (1.105) then takes on the desired form

$$r_{ij}'(x) = \left\{ g_j(x) + \sum_{m=1}^{M} g_m(x) \; r_{mj}(x) \right\}$$

$$\cdot \left\{ h_i(x) + \sum_{m=1}^{M} h_m(x) \; r_{im}(x) \right\} ,$$
(1.107)

for the system of M^2 equations for $r_{ij}(x)$, $i = 1, 2, \ldots,$ M; $j = 1, 2, \ldots,$ M. The initial conditions are

$$r_{ij}(0) = 0 .$$
(1.108)

The initial value problem for $u(t,x)$ is given by
the system of $M^2 + M$ ordinary differential equations and
initial conditions,

$$r'_{ij}(x) = \left\{ g_j(x) + \sum_{m=1}^{M} g_m(x)\, r_{mj}(x) \right\}$$

(1.109)

$$\cdot \left\{ h_i(x) + \sum_{m=1}^{M} h_m(x)\, r_{im}(x) \right\},$$

$$c'_i(x) = \left\{ f(x) + \sum_{m=1}^{M} g_m(x)\, c_m(x) \right\}$$

(1.110)

$$\cdot \left\{ h_i(x) + \sum_{m=1}^{M} h_m(x)\, r_{im}(x) \right\},$$

$$r_{ij}(0) = 0,$$

(1.111)

$$c_i(0) = 0,$$

(1.112)

$$i,j=1,2,\ldots,M,$$

and the representation

$$u(t,x) = f(t) + \sum_{m=1}^{M} g_m(t)\, c_m(x),$$

(1.113)

$$0 \le t \le x.$$

To solve the integral Eq. (1.82) for a given interval length
$x = a \le x$, one integrates Eqs. (1.109) and (1.110) subject to
Eqs. (1.111) and (1.112) from $x = 0$ to $x = a$. At the com-
pletion of the integration, the function $u(t,a) = u(t)$ is
evaluated using Eq. (1.113).

1.5   EIGENVALUES AND CRITICAL LENGTHS

Consider the homogeneous integral equation

$$\phi(t) = \lambda \int_0^x e^{t+y} \phi(y) \; dy \; , \qquad (1.114)$$

$$0 \leq t \leq x.$$

It is clear that

$$\phi(t) \equiv 0, \qquad (1.115)$$

$$0 \leq t \leq x,$$

is a solution, the so-called trivial solution. But generally this is of no interest. We wish to determine for what interval lengths (if any) there is a nontrivial solution. It is clear that if ϕ is a nontrivial solution, then $k\phi$ is also a nontrivial solution, where k is a nonzero constant.

We can find a relation between the parameter λ and the interval length x for which Eq. (1.114) has a nontrivial solution. Rearranging the integral in Eq. (1.114), we see that

$$\phi(t) = \lambda \; e^t \int_0^x e^y \; \phi(y) \; dy \; . \qquad (1.116)$$

Setting

$$m(x) = \int_0^x e^y \; \phi(y) \; dy \; , \qquad (1.117)$$

this becomes

$$\phi(t) = \lambda e^t m(x) \, , \tag{1.118}$$

$$0 \leq t \leq x.$$

But then use of Eq. (1.118) in Eq. (1.117) yields

$$m(x) = \int_0^x e^y \lambda e^y m(x) \, dy \, . \tag{1.119}$$

Since

$$m(x) \neq 0 \, , \tag{1.120}$$

we have

$$1 = \lambda \int_0^x e^{2y} \, dy \tag{1.121}$$

or

$$\lambda = 2/(e^{2x} - 1) \, , \tag{1.122}$$

which gives the desired relation between λ and the interval length x for which the homogeneous integral equation
has a nontrivial solution.

For $\lambda = 1$ we find that the critical length is

$$x_1 = (1/2) \ln 3 \, . \tag{1.123}$$

The homogeneous equation has a nontrivial solution for precisely that interval length for which the inhomogeneous equation fails to have a unique solution (see Eq. (1.26)).

This is the content of Fredholm's famous alternative theorem which holds for quite general kernels. It follows that we can determine where the homogeneous equation has a nontrivial solution by using the initial value method to determine values of the interval length x for which the solution of the inhomogeneous equation becomes unbounded. In effect, we can determine $x = x(\lambda)$; the inverse relationship $\lambda = \lambda(x)$ gives a value of the parameter λ for each x for which the homogeneous equation has a nontrivial solution. Such values of λ are called eigenvalues. The nontrivial solutions are called eigenfunctions.

We give numerical results which show that the solutions become unbounded as x gets closer to the critical length, for a given λ. We also give an example of estimating the relationship, $x = x(\lambda)$.

We also observe that as x tends to a critical length, a solution of an inhomogeneous integral equation approaches a nontrivial solution of the homogeneous integral equation. See Eqs. (1.118) and (1.34).

1.6   NUMERICAL RESULTS

A Fortran program has been written to solve the initial value problem described in the previous section and, hence, the integral equation Eq. (1.82). The results of four numerical experiments are described in this section.

First consider the kernel to be

Table 1.1

Numerical Results for the Kernel e^{-t-y}

a. Interval Length x = 1.0		
t	u(t,x) Initial Value Method	u(t,x) Exact
0	2.1135387	2.1135399
0.1	2.0075715	2.0075726
0.2	1.9116883	1.9116894
0.3	1.8249297	1.8249306
0.4	1.7464273	1.7464281
0.5	1.6753953	1.6753961
0.6	1.6111230	1.6111237
0.7	1.5529669	1.5529675
0.8	1.5003452	1.5003457
0.9	1.4527310	1.4527315
1.0	1.4096480	1.4096484
b. Interval Length x = 2.0		
t	u(t,x) Initial Value Method	u(t,x) Exact
0	2.6982213	2.6982254
0.2	2.3903860	2.3903893
0.4	2.1383518	2.1383545
0.6	1.9320036	1.9320058
0.8	1.7630600	1.7630618
1.0	1.6247407	1.6247422
1.2	1.5114944	1.5114956
1.4	1.4187762	1.4187772
1.6	1.3428650	1.3428658
1.8	1.2807141	1.2807148
2.0	1.2298293	1.2298298

$$k(t,y) = e^{-t} e^{-y} , \qquad\qquad (1.124)$$

and

$$f(t) = 1 . \qquad\qquad (1.125)$$

The closed-form solution found in Section 1.2 is

$$u(t,x) = 1 + c(x) e^{-t} , \qquad\qquad (1.126)$$

$$0 \le t \le x,$$

where

$$c(x) = (2 - 2e^{-x})/(1 + e^{-2x}) , \qquad\qquad (1.127)$$

$$0 \le x < \infty.$$

The results of the initial value calculations using a step size of .0025 for x = 1 and x = 2, using a fourth-order Runge-Kutta/Adams-Moulton method are shown in Table 1.1. The calculation is based on Eqs. (1.109)-(1.113).

In the second example the kernel is

$$k(t,y) = e^{t+y} = e^t e^y ,$$ (1.128)

and the forcing function is unchanged. This differs from
the first example in that now $c(x)$ becomes infinite for a
finite value of x. As we have seen in Section 1.2, this
value of x is the positive root of the equation

$$3 - e^{2x} = 0$$ (1.129)

which is

$$x_{crit} = (1/2) \ln 3 = 0.54930615 .$$ (1.130)

The calculation is performed for two interval lengths,
$x = .95 \, x_{crit} = .52184$ with step size of .00031, and
$x = .99_{crit} = .54381$ with step size of .00016. The exact
solution is

$$u(t,x) = 1 + c(x) e^t , \qquad\qquad 0 \le t \le x,$$ (1.131)

$$c(x) = (2e^x - 2)/(3 - e^{2x}) , \qquad 0 \le x < x_{crit}.$$ (1.132)

Results are shown in Table 1.2. We note that the solution,
$u(t,x)$, becomes larger in magnitude the closer we are to the
critical length because the denominator in Eq. (1.132) is tend-
ing to zero. Still, the initial value method produces very
good results. Later, in Example 4, we examine the problem
of estimating numerically just where this unboundedness
occurs.

Table 1.2

Numerical Results for the Kernel e^{t+y}

a. Interval Length x = 0.52184		
t	u(t,x) Initial Value Method	u(t,x) Exact
0	9.5563613	9.5549738
0.05218408	10.014723	10.013261
0.10436817	10.497638	10.496098
0.15655225	11.006424	11.004801
0.20873634	11.542465	11.540755
0.26092042	12.107221	12.105420
0.31310450	12.702231	12.700334
0.36528859	13.329116	13.327116
0.41747268	13.989582	13.987476
0.46965677	14.685430	14.683211
0.52184084	15.418554	15.416216
b. Interval Length x = 0.54381		
t	u(t,x) Initial Value Method	u(t,x) Exact
0	44.578135	44.583903
0.05438131	47.013592	47.019683
0.10876262	49.585161	49.591592
0.16314393	52.300446	52.307237
0.21752523	55.167482	55.174652
0.27190654	58.194748	58.202318
0.32628785	61.391199	61.399193
0.38066917	64.766291	64.774731
0.43505047	68.330006	68.338919
0.48943177	72.092887	72.102299
0.54381308	76.066066	76.076003

In the third example the integral equation is

$$u(t) = e^t - t + \int_0^x t(1-e^{ty}) \, u(y) \, dy \, , \tag{1.133}$$

$$0 \le t \le 1 \, ,$$

an equation with a nondegenerate kernel. This is to be solved by approximating the kernel by the polynomial

$$K_M(t,y) = -t \left[ty + \frac{(ty)^2}{2!} + \dots + \frac{(ty)^M}{M!} \right], \tag{1.134}$$

for $M = 2, 3$, and 4. The exact solution is

$$u(t) = 1 \, , \tag{1.135}$$

$$0 \le t \le 1;$$

and the results of the computations using the initial-value method with step size of .005 are displayed in Table 1.3 They show that the approximation improves as M increases from 2 to 4, but the results for $M = 2$ are quite accurate in themselves.

The fourth example shows that eigenvalues or critical lengths may be obtained. The kernel is considered to be

$$k(t,y) = \lambda \, e^{t+y} \, , \tag{1.136}$$

and

$$f = 0 \, . \tag{1.137}$$

Table 1.3

Results for the Third Example

t	$u(t,x)$ M=2	$u(t,x)$ M=3	$u(t,x)$ M=4
0	1.0000000	1.0000000	1.0000000
0.1	.99993922	.99998941	.99999848
0.2	.99979839	.99995811	.99999374
0.3	.99972437	.99991608	.99998606
0.4	.99998641	.99989542	.99997808
0.5	1.0009889	.99996302	.99997751
0.6	1.0032858	1.0002349	1.0000014
0.7	1.0075960	1.0008919	1.0000820
0.8	1.0148211	1.0021971	1.0002739
0.9	1.0260644	1.0083320	1.0006632
1.0	1.0426519	1.0013792	1.0013792

Then for various values of λ the differential equations for r and c are integrated with step size .005 until r' becomes sufficiently large. The exact relation from section 1.5 is

$$\lambda = \frac{2}{e^{2x_{crit}} - 1} . \qquad (1.138)$$

The results are shown graphically in Figure 1.1. Nonlinear extrapolation techniques are capable of producing extremely accurate results. Details will be given subsequently.

These calculations were carried out by B. Vereeke.

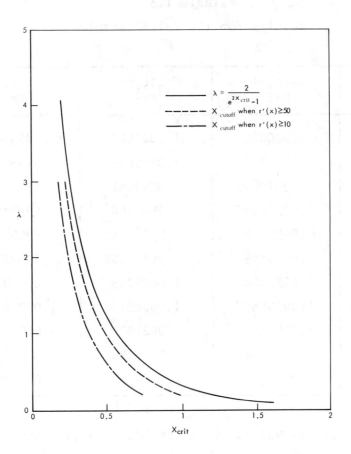

Figure 1.1   The First Eigenvalue as a Function of Interval
             Length

1.7   SUFFICIENCY OF CAUCHY SYSTEM

It is to be verified that $u(t,x)$ as determined by the Cauchy problem of Eqs. (1.109)-(1.113) of Section 3 does satisfy the integral Eq. (1.88) where the kernel is given by Eq. (1.83). In other words, it is to be shown that

$$u(t,x) = f(t) + \int_0^x \sum_{m=1}^M g_m(t) \, h_m(y) \, u(y,x) \, dy \, , \quad (1.139)$$

or, since $u(t,x)$ is given by Eq. (1.113) in terms of the coefficients $c_i(x)$, that

$$\sum_{m=1}^M g_m(t) \, c_m(x)$$

$$= \sum_{m=1}^M g_m(t) \left\{ \int_0^x h_m(y) \, f(y) \, dy \right.$$

$$\left. + \sum_{n=1}^M c_n(x) \int_0^x h_m(y) \, g_n(y) \, dy \right\} \, . \qquad (1.140)$$

The above equation will be satisfied if

$$c_i(x) = \int_0^x h_i(y) \, f(y) \, dy + \sum_{n=1}^M c_n(x) \int_0^x h_i(y) \, g_n(y) \, dy \, ,$$

$$i=1,2,\ldots,M. \qquad (1.141)$$

The plan is to establish relation (1.141) and hence Eqs. (1.140) and (1.139).

For ease of notation, let us define

$$F_i(x) = \int_0^x h_i(y) \, f(y) \, dy \, ,$$

$$G_{in}(x) = \int_0^x h_i(y) \, g_n(y) \, dy \, , \qquad\qquad (1.142)$$

$$i,n=1,2,\ldots,M.$$

Then Eq. (1.141) becomes

$$c_i(x) = F_i(x) + \sum_{n=1}^{M} G_{in}(x) \, c_n(x) \, , \qquad (1.143)$$

In order to show that relation (1.141) is fulfilled, we will establish that the right-hand side of Eq. (1.143) satisfies the same initial value problem as does the left-hand side. As we know, the left-hand side satisfies Eqs. (1.109)-(1.112).

Define the function $R_i(x)$ to be the right-hand side, namely,

$$R_i(x) = F_i(x) + \sum_{n=1}^{M} G_{in}(x) \, c_n(x) \, , \qquad (1.144)$$

$$i=1,2,\ldots,M.$$

In what follows, we do not explicitly indicate the x-dependence of the various functions, since this is to be understood. We wish to derive a differential equation for $R_i = R_i(x)$. By differentiation of Eq. (1.144), we have

$$R'_i = h_i f + \sum_{m=1}^{M} h_i \, g_m \, c_m$$

$$+ \sum_{n=1}^{M} G_{in} \left[h_n + \sum_{p=1}^{M} r_{np} \, h_p \right] \left[f + \sum_{m=1}^{M} c_m \, g_m \right]$$

$$= \left\{ f + \sum_{m=1}^{M} g_m \, c_m \right\} \left\{ h_i + \sum_{n=1}^{M} G_{in} \left[h_n + \sum_{p=1}^{M} r_{np} \, h_p \right] \right\}$$

$$= \left\{ f + \sum_{m=1}^{M} g_m \, c_m \right\}$$

$$\cdot \left\{ h_i + \sum_{n=1}^{M} h_n \left[G_{in} + \sum_{p=1}^{M} G_{ip} \, r_{pn} \right] \right\} . \tag{1.145}$$

Define $Q_{ij}(x)$ to be

$$Q_{ij}(x) = G_{ij}(x) + \sum_{p=1}^{M} G_{ip}(x) \, r_{pj}(x) , \tag{1.146}$$

$$i,j=1,2,\ldots,M.$$

Then

$$R'_i = \left[f + \sum_{m=1}^{M} g_m \, c_m \right] \left[h_i + \sum_{n=1}^{M} h_n \, Q_{in} \right] \tag{1.147}$$

and

$$R_i(0) = 0 . \tag{1.148}$$

Note that $R_i(x)$ will be equal to $c_i(x)$ if $Q_{ij}(x) = r_{ij}(x)$ $i,j,=1,2,\ldots,M$.

What we know about $r_{ij}(x)$ is that it satisfies Eqs. (1.109) and (1.111). Let us derive analogous equations for $Q_{ij}(x)$. Differentiating both sides of Eq. (1.146) we have

$$Q'_{ij} = h_i \, g_j + \sum_{p=1}^{M} h_i \, g_p \, r_j$$

$$+ \sum_{p=1}^{M} G_{ip}\left[h_p + \sum_{m=1}^{M} h_m \, r_{pm}\right]\left[g_j + \sum_{m=1}^{M} g_m \, r_{mj}\right]$$

$$= h_i\left[g_j + \sum_{p=1}^{M} g_p \, r_{pj}\right] + \left[\sum_{p=1}^{M} G_{ip} \, h_p\right.$$

$$+ \sum_{p=1}^{M} h_p \sum_{n=1}^{M} G_{in} \, r_{np}\left]\left[g_j + \sum_{m=1}^{M} g_m \, r_{mj}\right]\right.$$

$$= h_i\left[g_j + \sum_{m=1}^{M} g_m \, r_{mj}\right] + \sum_{p=1}^{M} h_p\left[G_{ip} + \sum_{n=1}^{M} G_{in} \, r_{np}\right.$$

$$\cdot \left[g_j + \sum_{m=1}^{M} g_m \, r_{mj}\right] , \qquad\qquad (1.149)$$

or

$$Q'_{ij} = \left[g_j + \sum_{m=1}^{M} g_m \, r_{mj}\right]\left[h_i + \sum_{p=1}^{M} h_p \, Q_{ip}\right] . \quad (1.150)$$

Also we have

$$Q_{ij}(0) = 0 . \qquad\qquad\qquad (1.151)$$

Comparing Eqs. (1.109), (1.140), (1.111), and (1.141),
we see that

$$Q_{ij}(x) = r_{ij}(x) , \qquad (1.152)$$
$$i,j=1,2,\ldots,M,$$

and, therefore, that

$$c_i(x) = R_i(x) , \qquad (1.153)$$
$$i=1,2,\ldots,M,$$

according to standard uniqueness theorems from the theory
or ordinary differential equations.

Consequently, it is true that the functions $c_i(x)$,
$i = 1, 2, \ldots, M$, fulfill relation (1.141), which implies
that $u(t,x)$ fulfills the integral Eq. (1.139). This completes the proof of the sufficiency of the Cauchy system for
lengths smaller than the first critical length. Various
methods are available for continuing the integration past
the critical lengths. Perhaps the simplest procedure is to
integrate in the complex x-plane, regarding x, r_{ij}, and
c_i as complex variables, on paths which avoid the critical
lengths.

1.8   RESOLVENTS

In an earlier section we considered the integral equation

$$u(t,x) = f(t) + \int_0^x e^{-(t+y)} u(y,x) \, dy \, , \qquad (1.154)$$
$$0 \leq t \leq x,$$

and we saw that the solution can be represented in the form

$$u(t,x) = f(t) + \int_0^x K(t,y,x) \, f(y) \, dy \, , \qquad (1.155)$$
$$0 \leq t \leq x,$$

where the resolvent kernel K is

$$K(t,y,x) = \frac{2e^{-(t+y)}}{1+e^{-2x}} \, , \qquad (1.156)$$
$$0 \leq t, y \leq x.$$

The resolvent K in no way involves the forcing function f, and the earlier equation shows the explicit linear dependence of the solution upon the forcing function. Let us now see that the resolvent K itself satisfies an integral equation. Write

$$\int_0^x K(t,y,x) \, f(y) \, dy = \int_0^x e^{-(t+y)}$$

$$\cdot \left[f(y) + \int_0^x K(y,y',x) \, f(y') \, dy' \right] dy$$

$$(1.157)$$

It follows that

$$\int_0^x K(t,y,x) \, f(y) \, dy$$

$$= \int_0^x \left[e^{-(t+y)} + \int_0^x e^{-(t+y')} \, K(y',y,x) \, dy' \right] f(y) \, dy \; ,$$

$$(1.158)$$

a relation which must hold for arbitrary functions f. But this implies that

$$K(t,y,x) = e^{-(t+y)} + \int_0^x e^{-(t+y')} \, K(y',y,x) \, dy' \; ,$$

$$(1.159)$$

which is the desired integral equation for the resolvent K. We can express the resolvent K in terms of the auxiliary functions J and r :

$$K(t,y,x) = e^{-y} J(t,x) = e^{-y} \left[e^{-t} + e^{-t} \, r(x) \right] \; .$$

$$(1.160)$$

It is possible to obtain a similar representation for the resolvent of a general degenerate kernel. To do this we observe that the resolvent K satisfies the integral equation

$$K(t,y,x) = k(t,y) + \int_0^x k(t,y') \, K(y',y,x) \, dy' \; ,$$

$$(1.161)$$

or

$$K(t,y,x) = \sum_{m=1}^{M} g_m(t) \, h_m(y) + \int_0^x k(t,y') \, K(y'y,x) \, dy' \, .$$

(1.162)

From this it follows that

$$K(t,y,x) = \sum_{m=1}^{M} J_m(t,x) \, h_m(y) \, ,$$

(1.163)

which is the desired representation in terms of the auxiliary functions J_1, J_2, ..., J_M. Also we see that

$$K(t,y,x) = \sum_{m=1}^{M} \left[g_m(t) + \sum_{n=1}^{M} g_n(t) \, r_{nm}(x) \right] h_m(y)$$

$$= k(t,y) + \sum_{n=1}^{M} \sum_{m=1}^{M} g_n(t) \, r_{nm}(x) \, h_m(y) \, ,$$

(1.164)

which is the desired representation for the resolvent in term of the auxiliary functions r_{11}, r_{12}, ... r_{MM}.

This result may be used computationally. We need not compute the functions c_1, c_2, ..., c_M. We need only compute the M^2 functions r_{11}, r_{12}, ..., r_{MM}. Then we express the resolvent K in terms of these functions, and finally we express u as

$$u(t,x) = f(t) + \int_0^x K(t,y,x) \, f(y) \, dy \, ,$$

(1.65)

the integral being evaluated by means of a quadrature
formula.

Lastly, we notice that the auxiliary function Φ, solution of the integral equation

$$\Phi(t,x) = k(t,x) + \int_0^x k(t,y') \; \Phi(y',x) \; dy' \; , \quad (1.166)$$

$$0 \leq t \leq x,$$

is

$$\Phi(t,x) = K(t,x,x) \; . \tag{1.167}$$

This implies that the function Φ can also be expressed in terms of the functions J_1, J_2, ..., J_M or r_{11}, r_{12}, ..., r_{MM}.

From Eqs. (1.146) and (1.152), it is seen that the functions $r_{ij}(x)$ satisfy the linear algebraic equations

$$r_{ij}(x) = G_{ij}(x) + \sum_{p=1}^{M} G_{ip}(x) \; r_{pj}(x) \; . \tag{1.168}$$

For a fixed index j (and a particular x) this is a system of M equations for r_{ij}, r_{2j}, ..., r_{Mj} .

The initial value method can be used in conjunction with the algebraic method. Suppose that Eqs. (1.168) are solved for all of the values of the r_{ij} for some x, and Eqs. (1.197) solved for the coefficients c_i. Then these values may be employed as known initial conditions, and the differential systems (1.109) and (1.111) will enable us to study r_{ij} and c_i as functions of x.

1.9   SUMMARY

We have both derived and validated an initial value problem for the solution of Fredholm integral equations with degenerate kernels. We have tested the method on several simple examples for which analytical solutions could be found. We also tested the method numerically, and we found that it produces fine results. We also saw that, in some cases, there are interval lengths for which the solution of the inhomogeneous integral equation becomes unbounded. At these lengths, the homogeneous integral equation has a non-trivial solution, which we can also determine.

Now we are ready to treat more complex integral equations by imbedding methods.

EXERCISES

1. Work out the imbedding method for the kernel

    $$k(t,y) = e^{t+y} ,$$

    and repeat the calculations. Compare against the known
    solution.

2. Derive Eq. (1.138).

3. Consider the kernel

    $$k(t,y) = e^{t+y} + 2t .$$

    Find the critical lengths and corresponding eigenfunc-
    tions.

4. Consider the problem of integrating past the first
    critical length.

5. Give the derivation of the complete initial value problem
    beginning with the system of linear algebraic equations
    (1.87) for $c_i = c_i(x)$, and for an appropriate set of
    auxiliary functions. Hint: Use vector-matrix notation.

6. Derive the complete initial value problem for the kernel

    $$k(t,y) = \int_0^1 g(t,z) \, h(y,z) \, dz .$$

REFERENCES

1. H. Kagiwada, R. Kalaba and B. Vereeke, "The Invariant
    Imbedding Numerical Method for Fredholm Integral
    Equations with Degenerate Kernels," *Journal of Ap-
    proximation Theory,* Vol. 1, No. 1 (1968), pp. 355-364.

2. R. Kalaba and M. Scott, "An Initial Value Method for
    Fredholm Integral Equations with Generalized Degenerate
    Kernels," *Journal of Approximation Theory,* to appear.

CHAPTER 2

SEMIDEGENERATE KERNELS

2.1   INTRODUCTION

Consider the family of integral equations

$$u(t,x) = g(t) + \int_0^x k(t,y)\, u(y,x)\, dy \, , \qquad (2.1)$$

$$0 \le t \le x \, ,$$

where the kernel k and the inhomogeneous term g are given, and the function u is to be determined for $0 \le t \le x$. The upper limit has been written as x, for we intend to study the solution u primarily as a function of x for a fixed value of t, $x \ge t$. This also accounts for the fact that u is written as u(t,x), rather than u(t), in Eq. (2.1).

We assume that the kernel has the form

$$k(t,y) = \begin{cases} \displaystyle\sum_{i=1}^{M} \alpha_i(t)\beta_i(y) \,, & 0 \leq y \leq t, \\[2em] \displaystyle\sum_{j=1}^{N} \gamma_j(t)\delta_j(y) \,, & t < y \leq x. \end{cases} \qquad (2.2)$$

An important case is that for which

$$k(t,y) = \sum_{\ell=1}^{L} c_\ell\, e^{-\lambda_\ell |t-y|} . \qquad (2.3)$$

Notice that this is a special case of Eq. (2.2) for which

$$k(t,y) = \begin{cases} \displaystyle\sum_{\ell=1}^{L} c_\ell\, e^{-\lambda_\ell (t-y)} \,, & 0 \leq y \leq t, \\[2em] \displaystyle\sum_{\ell=1}^{L} c_\ell\, e^{-\lambda_\ell (y-t)} \,, & t < y \leq x. \end{cases} \qquad (2.4)$$

In the event that the kernel k is not given in the form displayed in Eq. (2.2), it may be possible to approximate it by an appropriate series, e.g., a sum of powers, or Legendre polynomials, or a trigonometric series.

Our aim is to transform the given integral equation into an initial value problem, i.e., a system of ordinary differential equations with known initial conditions. Modern digital, analog and hybrid computers can solve several thousand such simultaneous differential equations with speed and accuracy.

We will see that the complete initial value problem involves a differential equation for u evaluated at a fixed point t, as well as the auxiliary functions $e_m(x)$ and $r_{mn}(x)$, m=1,2,...,M;   n=1,2,...,N.

2.2   DERIVATION OF THE CAUCHY SYSTEM

Differentiate Eq. (2.1) with respect to x. This yields

$$u_x(t,x) = k(t,x)\, u(x,x) + \int_0^x k(t,y)\, u_x(y,x)\, dy,$$

$$0 \le t \le x .$$

(2.5)

Next introduce the auxiliary function Φ as the solution of the integral equation

$$\Phi(t,x) = k(t,x) + \int_0^x k(t,y)\, \Phi(y,x)\, dy ,$$

$$0 \le t \le x .$$

(2.6)

Equation (2.5) may now be regarded as an integral equation for the function u_x. Its solution is

$$u_x(t,x) = \Phi(t,x)\, u(x,x) ,$$

$$0 \le t \le x .$$

(2.7)

This is one of the basic differential equations. To use it, we must consider the two factors on the right hand side. $\Phi(t,x)$ and $u(x,x)$. We begin by considering $\Phi(t,x)$.

The defining Eq. (2.6) may be rewritten

$$\Phi(t,x) = \sum_{j=1}^{N} \gamma_j(t) \; \delta_j(x) + \int_0^x k(t,y) \; \Phi(y,x) \; dy \; . \quad (2.8)$$

This suggests introducing the new auxiliary functions J_n, $n = 1, 2, \ldots, N$, as solutions of the integral equations

$$J_n(t,x) = \gamma_n(t) + \int_0^x k(t,y) \; J_n(y,x) \; dy \; , \quad (2.9)$$

$$0 \leq t \leq x, \quad n=1,2,\ldots,N.$$

In terms of the functions $J_1, J_2, \ldots, J_N$, the function Φ is given as

$$\Phi(t,x) = \sum_{j=1}^{N} J_j(t,x) \; \delta_j(x) \; . \quad (2.10)$$

This follows from the principle of superposition for linear systems. We shall now obtain differential equations for the functions $J_1, J_2, \ldots, J_N$.

Differentiate Eq. (2.9) with respect to x. The result is

$$J_n'(t,x) = k(t,x) \; J_n(x,x) + \int_0^x k(t,y) \; J_n'(y,x) \; dy \; ,$$

$$0 \leq t \leq x \; . \quad (2.11)$$

Regard Eq. (2.11) as an integral equation for the function J_n'. Its solution is

$$J_n'(t,x) = \Phi(t,x)\ J_n(x,x)\ , \qquad\qquad (2.12)$$

$$0 \leq t \leq x, \quad n=1,2,\ldots,N.$$

Recall that the function Φ is expressed in terms of the functions J_1, J_2, ..., J_N by means of Eq. (2.10). We now turn our attention to the functions $J_1(x,x)$, $J_2(x,x)$, ..., $J_N(x,x)$.

According to Eq. (2.9) we may write

$$J_n(x,x) = \gamma_n(x) + \int_0^x k(x,y)\ J_n(y,x)\ dy\ , \qquad (2.13)$$

$$x \geq 0,\ n=1,2,\ldots,N.$$

This equation may be rewritten in the form

$$J_n(x,x) = \gamma_n(x) + \int_0^x \sum_{i=1}^M \alpha_i(x)\ \beta_i(y)\ J_n(y,x)\ dy$$

$$\qquad\qquad (2.14)$$

$$= \gamma_n(x) + \sum_{i=1}^M \alpha_i(x)\int_0^x \beta_i(y)\ J_n(y,x)\ dy\ .$$

By introducing the new dependent variables $r_{mn}(x)$, for $m = 1, 2, \ldots, M$; $n = 1, 2, \ldots, N$; $x \geq 0$,

$$r_{mn}(x) = \int_0^x \beta_m(y)\ J_n(y,x)\ dy\ , \qquad\qquad (2.15)$$

we finally obtain

Framingham State College
Framingham, Massachusetts

$$J_n(x,x) = \gamma_n(x) + \sum_{i=1}^{M} \alpha_i(x)\, r_{in}(x)\ , \qquad (2.16)$$

$$x \geq 0, \quad n=1,2,\ldots,N.$$

This part of the discussion is completed by obtaining a differential equation for the function r_{mn}, $m = 1, 2,$ ..., M; $n = 1, 2, \ldots$, N. Differentiate both sides of Eq. (2.15). This yields

$$r'_{mn}(x) = \beta_m(x)\, J_n(x,x) + \int_0^x \beta_m(y)\, J'_n(y,x)\ dy \qquad (2.17)$$

or

$$r'_{mn}(x) = \beta_m(x)\, J_n(x,x)$$
$$\qquad\qquad (2.18)$$
$$+ \int_0^x \beta_m(y)\, \Phi(y,x)\, J_n(x,x)\ dy\ .$$

This becomes

$$r'_{mn}(x) = J_n(x,x) \left\{ \beta_m(x) + \int_0^x \beta_m(y)\, \Phi(y,x)\ dy \right\}\ . \qquad (2.19)$$

To simplify still further we evaluate the integral in the last equation as follows:

$$\int_0^x \beta_m(y) \; \Phi(y,x) \; dy = \int_0^x \beta_m(y) \sum_{j=1}^{N} J_j(y,x) \; \delta_j(x)$$

$$= \sum_{j=1}^{N} \delta_j(x) \int_0^x \beta_m(y) \; J_j(y,x) \; dy$$

$$(2.20)$$

$$= \sum_{j=1}^{N} r_{mj}(x) \; \delta_j(x) \; .$$

Equation (2.19) then becomes the Riccati system

$$r_{mn}'(x) = \left[\gamma_n(x) + \sum_{i=1}^{M} \alpha_i(x) \; r_{in}(x) \right]$$

$$(2.21)$$

$$\cdot \left[\beta_m(x) + \sum_{j=1}^{N} r_{mj}(x) \; \delta_j(x) \right] \; .$$

$$m=1,2,\ldots,M; \quad n=1,2,\ldots,N. \quad x \geq 0.$$

Furthermore the initial conditions follow readily from the definition in Eq. (2.15),

$$r_{mn}(0) = 0 \; , \qquad\qquad\qquad\qquad (2.22)$$

$$m=1,2,\ldots,M; \quad n=1,2,\ldots,N.$$

The differential equation for the function J_n, for $x \geq t$, is Eq. (2.12), where the function Φ is given by Eq. (2.10). The initial condition on the function J_n at $x = t$ is

$$J_n(t,t) = \gamma_n(t) + \sum_{i=1}^{M} \alpha_i(t)\ r_{in}(t), \qquad (2.23)$$

$$n=1,2,\ldots,N; \quad 0 \le t \le x \ .$$

Our discussion of the first factor on the right hand side of Eq. (2.7) is now complete. We now turn to the second factor, $u(x,x)$.

According to the integral equation (2.1), we may write

$$u(x,x) = g(x) + \int_0^x k(x,y)\ u(y,x)\ dy$$

$$= g(x) + \int_0^x \sum_{i=1}^{M} \alpha_i(x)\ \beta_i(y)\ u(y,x)\ dy$$

$$\qquad\qquad\qquad\qquad\qquad\qquad\qquad (2.24)$$

$$= g(x) + \sum_{i=1}^{M} \alpha_i(x) \int_0^x \beta_i(y)\ u(y,x)\ dy$$

$$= g(x) + \sum_{i=1}^{M} \alpha_i(x)\ e_i(x) \ ,$$

where

$$e_m(x) = \int_0^x \beta_m(y)\ u(y,x)\ dy, \qquad (2.25)$$

$$m=1,2,\ldots,M; \quad x \ge 0 \ .$$

By differentiation we see that

$$e_m'(x) = \beta_m(x) \, u(x,x) + \int_0^x \beta_m(y) \, u_x(y,x) \, dy$$

$$= \beta_m(x) \, u(x,x) + \int_0^x \beta_m(y) \, \Phi(y,x) \, u(x,x) \, dy \tag{2.26}$$

$$= u(x,x) \left[\beta_m(x) + \int_0^x \beta_m(y) \, \Phi(y,x) \, dy \right] .$$

It follows with the aid of Eq. (2.20) that

$$e_m'(x) = u(x,x) \left[\beta_m(x) + \sum_{j=1}^{N} r_{mj}(x) \, \delta_j(x) \right] , \tag{2.27}$$

where use has been made of Eq. (2.20). The final differential equation for the function e_m is

$$e_m'(x) = \left[g(x) + \sum_{i=1}^{M} \alpha_i(x) \, e_i(x) \right]$$

$$\tag{2.28}$$

$$\cdot \left[\beta_m(x) + \sum_{j=1}^{N} r_{mj}(x) \, \delta_j(x) \right] ,$$

$$m=1,2,\ldots,M; \quad n=1,2,\ldots,N; \quad x \geq 0.$$

According to Eq. (2.25) the initial condition on the function e_m at $x = 0$ is

$$e_m(0) = 0 , \tag{2.29}$$

$$m=1,2,\ldots,M.$$

2.3   SUMMARY OF THE CAUCHY SYSTEM

Let us now summarize the Cauchy system. The functions $\{e_m\}$ and $\{r_{mn}\}$ are determined by the differential equations (2.28) and (2.21),

$$e_m'(x) = \left[g(x) + \sum_{i=1}^{M} \alpha_i(x) \, e_i(x) \right]$$

(2.28)

$$\cdot \left[\beta_m(x) + \sum_{j=1}^{N} r_{mj}(x) \, \delta_j(x) \right] ,$$

$$r_{mn}'(x) = \left[\gamma_n(x) + \sum_{i=1}^{M} \alpha_i(x) \, r_{in}(x) \right]$$

(2.21)

$$\cdot \left[\beta_m(x) + \sum_{j=1}^{N} r_{mj}(x) \, \delta_j(x) \right] ,$$

$$m=1,2,\ldots,M; \quad n=1,2,\ldots,N; \quad x \geq 0,$$

together with the initial conditions at $x = 0$ given in Eqs. (2.29) and (2.22),

$$e_m(0) = 0 ,$$

(2.29)

$$r_{mn}(0) = 0 ,$$

(2.22)

$$m=1,2,\ldots,M; \quad n=1,2,\ldots,N.$$

These equations are integrated from $x = 0$ to $x = t$. At $x = t$ the differential equations for u and J_n are adjoined; these are Eqs. (2.7) and (2.12) (recall that the function Φ is expressed in terms of $J_1, J_2, \ldots, J_N,$

in Eq. (2.10)):

$$u_x(t,x) = \Phi(t,x) \, u(x,x) \; , \tag{2.7}$$

$$J_n'(t,x) = \Phi(t,x) \, J_n(x,x) \; , \tag{2.12}$$

$$n=1,2,\ldots,N; \quad 0 \le t \le x \; ,$$

$$\Phi(t,x) = \sum_{j=1}^{N} J_j(t,x) \, \delta_j(x) \; , \tag{2.10}$$

and $u(x,x)$ and $J_n(x,x)$ are expressed in terms of e and r :

$$u(x,x) = g(x) + \sum_{i=1}^{M} \alpha_i(x) \, e_i(x) \; , \tag{2.24}$$

$$J_n(x,x) = \gamma_n(x) + \sum_{i=1}^{M} \alpha_i(x) \, r_{in}(x) \; . \tag{2.16}$$

The initial conditions on the functions u and J_n are given in Eqs. (2.24) and (2.16) for $x = t$; i.e.,

$$u(t,x) \, \big|_{x=t} = u(t,t) = g(t) + \sum_{i=1}^{M} \alpha_i(t) e_i(t), \tag{2.30}$$

and

$$J_n(t,x) \Big|_{x=t} = J_n(t,t)$$

$$(2.31)$$

$$= \gamma_n(t) + \sum_{i=1}^{M} \alpha_i(t)\, r_{in}(t)\ ,$$

$$n=1,2,\ldots,N; \quad t>0\ .$$

As was remarked earlier, the integral equation (2.1) might be given with a kernel of the form in Eq. (2.2). Or the given kernel might be approximated by such a semi-degenerate kernel. In some instances, though, it is better to develop a new theory based on the special form of the kernel. This is done in Chapter 4 for displacement kernels,

$$k(t,y) = k(|t-y|),\qquad\qquad(2.32)$$

degenerate kernels (Chapter 1),

$$k(t,y) = \sum_{i=1}^{L} \alpha_i(t)\, \beta_i(y) \qquad\qquad(2.33)$$

and in Chapter 5 for kernels of the form

$$k(t,y) = k_1(|t-y|) + k_2(|t+y|)\ .\qquad(2.34)$$

For these cases FORTRAN programs are already available. Furthermore systems of Fredholm integral equations may also be treated.

2.4   NUMERICAL EXAMPLES

The results of three numerical experiments run on the IBM 7044 will be presented. An integration step size of 0.005 is used throughout with an Adams-Moulton fourth-order integration scheme.

For Example 1 let

$$g(t) = 1 \qquad\qquad (2.35)$$

in Eq. (2.1), and let

$$k(t,y) = \begin{cases} \lambda y, & 0 \le y < t, \\ \lambda t, & t \le y \le x. \end{cases} \qquad (2.36)$$

Let

$$\lambda = 1, \quad x = 1. \qquad\qquad (2.37)$$

The closed-form solution is

$$u(t) = \cos t + \tan 1 \sin t. \qquad (2.38)$$

The results of both the initial value method calculations and the closed-form solution are given in Table 2.1. The execution time for the initial value calculation is less than 5 sec.

Table 2.1

Comparison of Numerical Results for u(t,x)   for Example 1

$\lambda = 1.0$, Interval Length = 1.0

t	Exact Values	Values Obtained Using Initial Value Method
0.1	1.15049	1.15048
0.2	1.28948	1.28947
0.3	1.41558	1.41558
0.4	1.52754	1.52754
0.5	1.62424	1.62424
0.6	1.70471	1.70471
0.7	1.76815	1.76815
0.8	1.81392	1.81392
0.9	1.84157	1.84157
1.0	1.85082	1.85082

Next consider eigenvalues or critical lengths. The differential equations for r and e are integrated for values of λ between 0.2 and 3.0 until r' becomes greater than 20.0 These cutoff values of x are shown in the Figure 2.1. The exact relation is

$$x_n = \frac{\pi/2 + (n-1)\pi}{\sqrt{\lambda}}, \qquad (2.39)$$

where x_n represents the nth critical length.

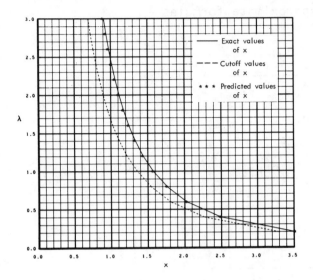

Figure 2.1 Eigenvalues for Example 1

By using nonlinear extrapolation techniques, extremely accurate critical lengths can be obtained. Let x_1 be the value of x at which r' becomes greater than 20.0. Also let

$$r_1 = r(x_1) \, , \qquad\qquad\qquad\qquad (2.40)$$

$$r_2 = r(x_2) \, , \qquad x_2 = x_1 - \Delta \, , \qquad\qquad (2.41)$$

and

$$r_3 = r(x_3) \, , \qquad x_3 = x_2 - \Delta \, . \qquad\qquad (2.42)$$

Then use the approximate formulas

$$r_1 = \frac{a_{-1}}{x_1 - x_{crit}} + a_o \; , \tag{2.43}$$

$$r_2 = \frac{a_{-1}}{x_2 - x_{crit}} + a_o \; , \tag{2.44}$$

and

$$r_3 = \frac{a_{-1}}{x_3 - x_{crit}} + a_o \; . \tag{2.45}$$

Solving for $x_{crit} = x_c$, the formula

$$x_c = \frac{x_3(x_2 - x_1) - x_2(x_3 - x_2)R}{(x_2 - x_1) - (x_3 - x_2)R} \tag{2.46}$$

is obtained, where

$$R = \frac{r_1 - r_2}{r_2 - r_3} \; . \tag{2.47}$$

The cutoff values, the extrapolated values, and the exact critical lengths are all shown in the Figure 2.1.

Example 2 has a forcing term of

$$g(t) = t \; , \tag{2.48}$$

and a kernel of

$$k(t,y) = \begin{cases} y(1-t), & 0 \le y \le t , \\ t(1-y), & t \le y \le x . \end{cases} \qquad (2.49)$$

The closed-form solution is

$$u(t) = \frac{\sin t}{\sin 1} , \qquad (2.50)$$

$$0 \le t \le 1 .$$

The kernel expressed in terms of Eq. (2.2) has both M and N equal to 1. It can also be thought of as

$$k(t,y) = \begin{cases} y - yt, & 0 \le y \le t , \\ t - ty , & t \le y \le x , \end{cases} \qquad (2.51)$$

with M and N equal to 2. When both forms of the kernel were calculated for comparison, the execution time for each was not more than 10 sec. The results are displayed in Table 2.2.

To determine the critical lengths, consider Example 2 with $g = 0$. Then u satisfies the boundary conditions

$$u(0) = 0 , \qquad (2.52)$$

and

$$u(x) = (x-1)\dot{u}(x) , \qquad (2.53)$$

and the differential equation

$$\ddot{u}(t) = \lambda u . \qquad (2.54)$$

Table 2.2

Comparison of Numerical Results for u(t) for Example 2

Interval Length = 1.0

t	Exact Values	Approximate Values M=2, N=2	Approximate Values M=1, N=1
0.1	0.118642	0.118641	0.118641
0.2	0.236098	0.236097	0.236097
0.3	0.351195	0.351194	0.351194
0.4	0.462783	0.462783	0.462783
0.5	0.569747	0.569747	0.569747
0.6	0.671018	0.671018	0.671018
0.7	0.765585	0.765585	0.765585
0.8	0.852502	0.852502	0.852502
0.9	0.930902	0.930902	0.930902
1.0	1.000000	1.000000	1.000000

Therefore,

$$u(t) = A \sin \sqrt{\lambda} t \, . \tag{2.55}$$

The relation between $\sqrt{\lambda}$ and the critical length x_c is

$$\sin \sqrt{\lambda} x_c = (x_c - 1) \sqrt{\lambda} \cos \sqrt{\lambda} x_c \, . \tag{2.56}$$

This equation can be solved by Newton's method. When $\lambda = 1$, it is found that

$$x_c = 4.42860 \, . \tag{2.57}$$

Using the initial value and extrapolation technique, it is found that

$$x_c = 4.42836 . \tag{2.58}$$

Example 3 concerns a clamped beam on an elastic foundation. The equilibrium configuration is described by a fourth-order differential equation,

$$u^{IV} = g(t) - ku(t) , \tag{2.59}$$

with boundary conditions

$$u(0) = 0 , \tag{2.60}$$

$$\dot{u}(0) = 0 , \tag{2.61}$$

$$u(1) = 0 , \tag{2.62}$$

and

$$\dot{u}(1) = 0 . \tag{2.63}$$

The fourth-order equation can be converted into an integral equation,

$$u(t) = \int_0^1 G(t,y) \; [g(y) - ku(y)] \; dy , \tag{2.64}$$

where the Green's function G on the interval $(0,1)$ has the form

$$G(t,y) = \begin{cases} -\dfrac{y^3}{6} + \dfrac{ty^2}{2} - t^2y^2 + \dfrac{t^2y^3}{2} + \dfrac{t^3y^2}{2} - \dfrac{t^3y^3}{3}, & y<t, \\[2em] -\dfrac{t^3}{6} + \dfrac{t^2y}{2} - t^2y^2 + \dfrac{t^3y^2}{2} + \dfrac{t^2y^3}{2} - \dfrac{t^3y^3}{3}, & y>t, \end{cases}$$

$$(2.65)$$

which is symmetric. Although the unexpanded version of this function is considerably more compact, the expanded form is advantageous to check the Fortran program. In terms of Eq. (2.2) for the kernel, it has M and N values

$$M = 6 \tag{2.66}$$

and

$$N = 6 . \tag{2.67}$$

If g(y) in Eq. (2.64) is 1, the problem becomes

$$u(t) = -k \int_0^1 G(t,y)u(y) \, dy + Q(t) , \tag{2.68}$$

where the forcing term is

$$Q(t) = \int_0^1 G(t,y) \, dy . \tag{2.69}$$

Expressed analytically, Eq. (2.69) is

$$Q(t) = \frac{t^2}{24} (1-t)^2 . \tag{2.70}$$

The initial value method calculations are carried out for both Eq. (2.70) and a quadrature evaluation of Eq. (2.69). The agreement is excellent. As a check, Eq. (2.59) is integrated by the method of complementary functions. The numerical results are shown in Table 2.3.

Table 2.3

Comparison of Numerical Results of $u(t,x)$ and $u(t)$ for Example 3

$$k = 1$$

t	Initial-Value Method for Computing $u(t,x)$ with Analytic Expression for Forcing Term	Complementary Function Method for Evaluation of $u(t)$ in the Fourth-Order Equation
0.0	0	
0.1	3.369×10^{-4}	3.37×10^{-4}
0.2	1.065×10^{-3}	1.06×10^{-3}
0.3	1.834×10^{-3}	1.83×10^{-3}
0.4	2.395×10^{-3}	2.40×10^{-3}
0.5	2.599×10^{-3}	2.60×10^{-3}
0.6	2.395×10^{-3}	2.40×10^{-3}
0.7	1.834×10^{-3}	1.83×10^{-3}
0.8	1.065×10^{-3}	1.06×10^{-3}
0.9	3.369×10^{-4}	3.37×10^{-4}
1.0	-7.276×10^{-12}	-9.31×10^{-10}

2.5   VOLTERRA INTEGRAL EQUATIONS

An interesting reduction in the initial value problem occurs in the event that the Volterra integral equation

$$u(t,x) = g(t) + \int_t^x k(t,y)u(y,x)\, dy , \qquad (2.71)$$

$$t<x ,$$

is to be considered. This implies that

$$\alpha_i(t) = 0, \qquad\qquad\qquad i=1,2,\ldots,N. \quad (2.72)$$

$$\beta_i(y) = 0 , \qquad\qquad\qquad i=1,2,\ldots,N. \quad (2.73)$$

The ordinary differential equations for $J_j(t,x)$ become

$$J_j'(t,x) = \gamma_j(x) \sum_{s=1}^N \delta_s(x) J_s(t,x) , \qquad (2.74)$$

$$j=1,2,\ldots,N, \quad x>t .$$

The initial condition on $J_j(t,x)$ at $x = t$ is

$$J_j(t,t) = \gamma_j(t), \qquad\qquad j=1,2,\ldots,N. \quad (2.75)$$

The differential equation for u is

$$u_x(t,x) = f(x) \sum_{j=1}^N \delta_j(x)\, J_j(t,x) , \qquad (2.76)$$

$$x>t ,$$

and the initial condition at $x = t$ is

$$u(t,t) = f(t) . \qquad\qquad (2.77)$$

The functions $\{r_{ij}\}$ and $\{e_j\}$ need not be computed. For $\{\gamma_j\}$ and $\{\delta_j\}$ continuous and bounded on an interval, standard theorems ensure the existence of solutions of Eqs. (2.74) and (2.76), and hence of the Volterra integral Eq. (2.71).

2.6   CRITICAL LENGTHS AND EIGENVALUES

Introduce a parameter λ into Eq. (2.1) by replacing $\alpha_i(t)$ and $\gamma_j(t)$ by $\lambda\alpha_i(t)$ and $\lambda\gamma_j(t)$, respectively. The new equations for $r_{ij}(x)$, formed from Eq. (2.21), may be integrated until the components of r_{ij} become very large. This provides a lower bound for the first critical length of each value of λ for which the integration is performed. To continue the integration beyond the critical length, we may solve the integral equations (2.9) for J_j, $j = 1, 2, \ldots,$ N, for a supercritical length, using any available method. For this length Eqs, (2.15) provide the new initial values for r_{ij}. This leads to estimates of the second critical lengths, etc. Alternatively, the Laurent expansions for the solutions r_{ij} in the neighborhood of a critical length may be employed.

2.7   FORTRAN PROGRAM

A program has been written in FORTRAN to solve integral equations with semidegenerate kernels. It produces
$u(t_i,x)$ for a given interval length x and a maximum of 100
different values of t_i .

First, the MN + M differential equations for r_{ij}
and e_i are integrated by an Adams-Moulton integration scheme
on the interval $0 \leq x \leq t_1$. When x reaches t_i, N + 1
additional differential equations for J_j and u are adjoined
and the integration continues. Each time that x reaches one
of the values t_i, N + 1 differential equations for J_j and
u with initial conditions at $x = t_i$ are added and integrated
until x equals the given interval length.

The user must provide a subroutine called KERNEL with
argument list (ALPHA, BETA, GAMMA, DELTA, X) and a function
subprogram called FORCE with a single argument (X). The subroutine KERNEL must dimension ALPHA, BETA, GAMMA, and DELTA
and evaluate ALPHA(I), BETA(I), GAMMA(J), and DELTA(J)
(I=1, M; J=1, N) at x . The function subprogram FORCE simply
evaluates the forcing term at x . The subroutine and function
subprogram used in Example 2 are shown at the end of the
program listing.

The input data are given on from 3 to 19 cards, de-
pending on the number of t_i specified. The first card is a
title, 76 alphameric characters in length, which is printed
at the top of the output for identification purposes. The
second card gives values for the variables NUMBER, M, N, STEP
and LENGTH. The variable NUMBER is the total number of t_i
to be input, the maximum being 100; STEP is the integration
step size to be used; LENGTH is the interval length. The

maximum values for M and N are 10. Beginning with the third card, the t_i's are given, six per card, in increasing order. See the listing for formats.

The output, headed by the title, repeats all of the input data, gives the interval length at which integration stopped (since the given length is not always an even multiple of the step size), and gives $u(t_i,x)$ for each t_i . The input cards and corresponding output for Example 2 are shown at the end of the listing.

```
$IBFTC SEMI
C
C     FREDHOLM INTEGRAL EQUATIONS WITH SEMIDEGENERATE KERNELS
C
C     ***************************************************************
      DIMENSION T(12100),R(10,10),TITLE(13)
      DIMENSION SMALLT(100),SUMR(10)
      COMMON /MARKS/ N,MID,MARK1,MARK2,KOUNTT,NUMBER
      COMMON /KERN/ ALPHA(10),BETA(10),DELTA(10),GAMMA(10)
      REAL LENGTH
C     ***************************************************************
C
C     TITLE IS 76 ALPHAMERIC CHARACTERS PRINTED AT TOP OF OUTPUT
C     NUMBER=TOTAL NUMBER OF T(I) TO BE INPUT
C     LENGTH=INTERVAL LENGTH
C     STEP=INTEGRATION STEP SIZE
C
      READ (5,5) (TITLE(I),I=1,13)
      READ (5,1) NUMBER,M,N,STEP,LENGTH
      READ (5,2) (SMALLT(I),I=1,NUMBER)
      X=0.0
      DO 50 I=1,12100
   50 T(I)=0.0
C
C     LOCATION MARKERS FOR STORAGE IN T
C               MARK1=LAST LOCATION FOR ORIGINAL DIFFERENTIAL EQUATIONS
C               MID  =LAST LOCATION FOR ALL DIFFERENTIAL EQUATIONS
C               MARK2=LAST LOCATION FOR DERIVATIVES OF ORIGINAL EQUATIONS
C               KOUNTT=THE NUMBER OF THE LAST T TO BE PASSED
C
      MARK1=M*N + M
      MID=MARK1
      MARK2=MID + M*N + M
      KOUNTT=0
      KK=KOUNTT + 1
      IF (SMALLT(1).EQ.0.0) GO TO 100
   60 CALL INT1(T,MID,X,STEP,M)
   70 CALL INT2(T,MID,X,STEP,M)
      KK=KOUNTT + 1
      IF (ABS(SMALLT(KK)-X).LE.STEP/2.0) GO TO 100
      IF (ABS(LENGTH-X).LE.STEP/2.0) GO TO 200
      GO TO 70
C
C     ZERO OUT DERIVATIVES
C
  100 X=SMALLT(KK)
      KOUNTT=KOUNTT + 1
      DO 110 I=1,MID
      J=MID + I
  110 T(J)=0.0
C
C     INITIAL CONDITIONS FOR N+1 NEW DIFFERENTIAL EQUATIONS
C
      CALL KERNEL(ALPHA,BETA,GAMMA,DELTA,X)
      SUME=0.0
      DO 120 J=1,N
      SUMR(J)=0.0
      DO 120 I=1,M
      K=(J-1)*M + I
```

```
  120 SUMR(J)=SUMR(J) + ALPHA(I)*T(K)
      DO 125 I=1,M
      K=M*N + I
  125 SUME=SUME + ALPHA(I)*T(K)
      DO 130 J=1,N
      KDIF=MID + J
  130 T(KDIF)=GAMMA(J) + SUMR(J)
      MID=MID + N + 1
      T(MID)=FORCE(X) + SUME
      MARK2=MID + M*N + M
      IF (KOUNTT.LT.NUMBER) GO TO 60
      IF (ABS(LENGTH-X).GT.STEP/2.0) GO TO 60
C
C     OUTPUT THE RESULTS
C
  200 WRITE (6,6) (TITLE(I),I=1,13)
      WRITE (6,3) LENGTH,X,STEP,M,N
      DO 210 I=1,NUMBER
      L=M*N + M + (N+1)*I
  210 WRITE (6,4) I,SMALLT(I),T(L)
C
    1 FORMAT (3I4,2E12.8)
    2 FORMAT (6E12.8)
    3 FORMAT (1H ,19X,30H**********INPUT DATA**********///1H ,24X,18HINT
     1ERVAL LENGTH = ,F7.4/1H ,22X,20HEND OF INTEGRATION= ,F7.4/1H ,19X,
     223HINTEGRATION STEP SIZE= ,F7.4/1H ,38X,4HM = ,I3/1H ,38X,4HN = ,I
     33//1H ,12X,44H**************************************************/////1H
     4,31X,7HRESULTS///1H ,18X,1HI,5X,4HT(I),14X,6HU(T,X)/)
    4 FORMAT (1H ,17X,I2,3X,F7.3,9X,1PE13.8)
    5 FORMAT (13A6)
    6 FORMAT (1H1,13A6//)
C
      CALL EXIT
      END
$IBFTC DAUX
C
C     DAUX FOR FREDHOLM INTEGRAL EQUATIONS WITH SEMIDEGENERATE KERNELS.
C
      SUBROUTINE DAUX (T,M,X,STEP)
C
C     ***************************************************************
      DIMENSION SUM1(10),SUM2(10),SUM4(100),T(12100)
      COMMON /MARKS/ N,MID,MARK1,MARK2,KOUNTT,NUMBER
      COMMON /KERN/ ALPHA(10),BETA(10),DELTA(10),GAMMA(10)
C     ***************************************************************
C
C     EVALUATE SUMS FOR R(I,J) AND E(I) DERIVATIVES.
C
      CALL KERNEL(ALPHA,BETA,GAMMA,DELTA,X)
      SUM3=0.0
      DO 20 I=1,10
      SUM1(I)=0.0
   20 SUM2(I)=0.0
      DO 30 I=1,M
      K=M*N + I
      SUM3=SUM3 + ALPHA(I)*T(K)
      DO 30 J=1,N
      KK=(J-1)*M + I
      SUM1(J)=SUM1(J) + T(KK) *ALPHA(I)
```

```
      SUM2(I)=SUM2(I) + T(KK) *DELTA(J)
   30 CONTINUE
C
C     EVALUATE SUM FOR CAPJ(J) AND U DERIVATIVES.
C
      IF (KOUNTT.EQ.0) GO TO 55
      DO 40 I=1,100
   40 SUM4(I)=0.0
      K=MARK1
      DO 50 I=1,KOUNTT
      DO 45 J=1,N
      L=K+J
   45 SUM4(I)=SUM4(I) + T(L)*DELTA(J)
      K=L+1
   50 CONTINUE
C
C     EVALUATE R(I,J) AND E(I) DERIVATIVES.
C
   55 DO 60 I=1,M
      KE=MID + M*N + I
      T(KE)=(FORCE(X) + SUM3) * (BETA(I) + SUM2(I))
      DO 60 J=1,N
      KR=MID + (J-1)*M + I
   60 T(KR)=(GAMMA(J) + SUM1(J)) * (BETA(I) + SUM2(I))
      IF (KOUNTT.EQ.0) GO TO 90
C
C     EVALUATE ALL OF THE CAPJ(J) AND U DERIVATIVES.
C
      K=MARK2
      DO 80 I=1,KOUNTT
      DO 70 J=1,N
      L=K+J
   70 T(L)=SUM4(I) * (GAMMA(J) + SUM1(J))
      K=L+1
      T(K)=(FORCE(X) + SUM3) * SUM4(I)
   80 CONTINUE
C
   90 RETURN
      END
$IBFTC INT1
      SUBROUTINE INT1(T,N,X,H,M)
      COMMON /NSEQ/ N2,N3,N4,N5,N6,N7,N8,N9,NN,KFLAG,IND
      COMMON /HSEQ/ H2,H4,H24,R
      DIMENSION T(1)
C
C     CALC. CONSTANTS TO BE USED IN PROGRAM
      NN=N
      N2=N*2
      N3=N*3
      N4=N*4
      N5=N*5
      N6=N*6
      N7=N*7
      N8=N*8
      N9=N*9
      H2=H*0.5
      H4=H2*0.5
      H24=H/24.0
      R=1.0/6.0
```

```
C       CALC. Y PRIME FOR INITIAL CONDITIONS
        CALL DAUX(T,M,X,H)
        DO 1 I=1,N
        N9I=I+N9
C       TEMPORARY STORAGE FOR Y
        N8I=I+N8
        NNI=I+NN
        T(N9I)=T(1)
C       STORE Y PRIME AT N-3 FOR USE IN A.M. INTEGRATION
      1 T(N8I)=T(NNI)
        KFLAG=0
        IND=0
        RETURN
        END
$IBFTC INT2
        SUBROUTINE INT2(T,N,X,H,M)
        COMMON /NSEQ/ N2,N3,N4,N5,N6,N7,N8,N9,NN,KFLAG,IND
        COMMON /HSEQ/ H2,H4,H24,R
        DIMENSION T(1)
        DOUBLE PRECISION Y,YP,YC
C
C       IND=FLAG FOR R.K. INTEGRATION WHEN EQUAL OR LESS THAN 3
C       KFLAG=FLAG FOR STORING PAST DERIVATIVES FOR USE IN A.M. INTEGRATION
C
C                  RUNGE-KUTTA INTEGRATION (6 STEPS AT H/2)
C       NOTE- TWO STEPS OF R.K. ARE DONE WITH EACH CALL TO THIS SUBR.,
C             SO THAT PRINTOUT POINTS WILL BE AT STEPS OF H.
C
        IND=IND+1
        KFLAG=KFLAG+1
        IF (IND.GT.3) GO TO 12
        DO 9 K=1,2
C               CALC. K1
        DO 1 I=1,N
        N2I=I+N2
        NNI=I+NN
      1 T(N2I)=T(NNI)*H2
C               CALC. K2
C       STEP UP X
        X=X+H4
        DO 2 I=1,N
        N9I=I+N9
        N2I=I+N2
      2 T(I)=T(N9I)+0.5*T(N2I)
        CALL DAUX(T,M,X,H)
C       STORE K2
        DO 3 I=1,N
        N3I=I+N3
        NNI=I+NN
      3 T(N3I)=T(NNI)*H2
C               CALC. K3
        DO 4 I=1,N
        N9I=I+N9
        N3I=I+N3
      4 T(I)=T(N9I)+0.5*T(N3I)
        CALL DAUX(T,M,X,H)
C       STORE K3
        DO 5 I=1,N
        N4I=I+N4
```

```
      NNI=I+NN
    5 T(N4I)=T(NNI)*H2
C            CALC. K4
C     STEP UP X
      X=X+H4
      DO 6 I=1,N
      N9I=I+N9
      N4I=I+N4
    6 T(I)=T(N9I)+T(N4I)
      CALL DAUX(T,M,X,H)
C     STORE K4
      DO 7 I=1,N
      N5I=I+N5
      NNI=I+NN
    7 T(N5I)=T(NNI)*H2
C     CALC. PREDICTED VALUE OF Y
      DO 8 I=1,N
      N9I=I+N9
      N2I=I+N2
      N3I=I+N3
      N4I=I+N4
      N5I=I+N5
      Y=T(N9I)+R*(T(N2I)+2.0*T(N3I)+2.0*T(N4I)+T(N5I))
C     STORE AS CURRENT VALUE OF Y
      T(I)=Y
C     STORE Y IN TEMPORARY STORAGE
    8 T(N9I)=T(I)
C     CALC. Y PRIME
      CALL DAUX(T,M,X,H)
C
C     STORE DERIVATIVES AS NEED FOR A.M.
      GO TO (10,11,9),KFLAG
C     STORE Y PRIME AT N-2
   10 DO 18 I=1,N
      N7I=I+N7
      NNI=I+NN
   18 T(N7I)=T(NNI)
      GO TO 9
C     STORE Y PRIME AT N-1
   11 DO 19 I=1,N
      N6I=I+N6
      NNI=I+NN
   19 T(N6I)=T(NNI)
    9 CONTINUE
      GO TO 17
C            ADAMS-MOULTON INTEGRATION
C
C     STORE CURRENT Y AND Y PRIME IN TEMPORARY STORAGE
   12 DO 13 I=1,N
      N9I=I+N9
      N2I=I+N2
      NNI=I+NN
      T(N9I)=T(I)
   13 T(N2I)=T(NNI)
C     CALC. PREDICTED VALUE OF Y
      DO 14 I=1,N
      N2I=I+N2
      N6I=I+N6
      N7I=I+N7
```

```
      N8I=I+N8
      N9I=I+N9
      YP=T(N9I)+H24*(55.0*T(N2I)-59.0*T(N6I)+37.0*T(N7I)-9.0*T(N8I))
C     STORE AS PREDICTED FUNCTIONAL VALUE OF Y AT X=X+H
   14 T(I)=YP
C     STEP UP X
      X=X+H
C     CALC. Y PRIME USING PREDICTED Y
      CALL DAUX(T,M,X,H)
C     CALC. CORRECTED Y
      DO 15 I=1,M
      NNI=I+NN
      N2I=I+N2
      N6I=I+N6
      N7I=I+N7
      N9I=I+N9
      YC=T(N9I)+H24*(9.0*T(NNI)+19.0*T(N2I)-5.0*T(N6I)+T(N7I))
C     STORE AS NEW CURRENT VALUE OF Y
   15 T(I)=YC
C     CALC. Y PRIME TO BE USED IN NEW STEP
      CALL DAUX(T,M,X,H)
C     REARRANGE STORAGE OF PREVIOUS DERIVATIVES.
      DO 16 I=1,N
      N8I=I+N8
      N7I=I+N7
      N6I=I+N6
      N2I=I+N2
C     Y PRIME (N-2) GOES TO (N-3)
      T(N8I)=T(N7I)
C     Y PRIME (N-1) GOES TO (N-2)
      T(N7I)=T(N6I)
C     Y PRIME (N) GOES TO (N-1)
   16 T(N6I)=T(N2I)
   17 RETURN
      END




$IBFTC KERNEL
C
C     FOR EX 2 WITH SEMIDEGENERATE KERNEL
C
      SUBROUTINE KERNEL(ALPHA,BETA,GAMMA,DELTA,X)
C
      DIMENSION ALPHA(2),BETA(2),DELTA(2),GAMMA(2)
      ALPHA(1)=1.0
      ALPHA(2)=-X
      BETA(1)=X
      BETA(2)=X
      GAMMA(1)=X
      GAMMA(2)=X
      DELTA(1)=1.0
      DELTA(2)=-X
C
      RETURN
      END
```

```
$IBFTC FORCE
C
      FUNCTION FORCE(X)
C
      FORCE=X
      RETURN
      END
$ENTRY          SEMI
   SAMPLE OUTPUT FOR SECOND EXAMPLE OF SEMIDEGENERATE KERNEL
   11   2   2   .005          1.0
  0.0          0.1          0.2          0.3          0.4          0.5
  0.6          0.7          0.8          0.9          1.0
```

SAMPLE OUTPUT FOR SECOND EXAMPLE OF SEMIDEGENERATE KERNEL

**********INPUT DATA**********

INTERVAL LENGTH =   1.0000
END OF INTEGRATION=   1.0000
INTEGRATION STEP SIZE=   0.0050
M =   2
N =   2

****

RESULTS

I	T(I)	U(T,X)
1	0.000	0.
2	0.100	1.18641E-01
3	0.200	2.36097E-01
4	0.300	3.51194E-01
5	0.400	4.62783E-01
6	0.500	5.69747E-01
7	0.600	6.71018E-01
8	0.700	7.65585E-01
9	0.800	8.52502E-01
10	0.900	9.30902E-01
11	1.000	1.00000E 00

2.8   DISCUSSION

We have considered Fredholm integral equations with semi-degenerate kernels, and have shown how to convert them to initial value problems. Numerical experiments have shown that the method works. A special form of these kernels lends to Volterra integral equations so these, too, can be treated as Cauchy problems. Critical lengths and eigenvalues may also be determined by solving Cauchy problems.

We leave the validation of the initial value method as an exercise.

A listing of a FORTRAN program for integral equations with semidegenerate kernels has been provided with sample input data and output.

EXERCISES

1. Validate the Cauchy system of Section 2.3.
2. Punch up the general computer program as listed in Section 2.7.
3. Use the program to solve numerical examples.
4. Determine critical lengths and eigenvalues.

REFERENCES

1. H. Kagiwada and R. Kalaba, "An Initial Value Theory for Fredholm Integral Equations with Semi-degenerate Kernels," *Journal of the Association for Computing Machinery*, Vol. 1 No. 3 (July 1970), pp. 412-419.
2. T. Kailath, "Fredholm Resolvents, Wiener-Hopf Equations

and Riccati Differential Equations,"*IEEE Transactions on Information Theory,* v. IT-15 (1969), pp. 665-672.

CHAPTER 3

RESOLVENTS FOR SEMIDEGENERATE KERNELS

●

3.1   INTRODUCTION

The solution $u(t)$ of the Fredholm integral equation

$$u(t) = g(t) + \int_0^x k(t,y)\ u(y)\ dy, \qquad (3.1)$$
$$0 \le t \le x\ ,$$

for arbitrary forcing function $g(t)$ may be expressed in terms of the resolvent kernel $K(t,y,x)$ as

$$u(t) = g(t) + \int_0^x K(t,y,x)\ g(y)\ dy\ . \qquad (3.2)$$

By substitution of this expression into the original integral equation, it is seen that the resolvent satisfies the Fredholm integral equation

$$K(t,y,x) = k(t,y) + \int_0^x k(t,z) \, K(z,y,x) \, dz \,, \quad (3.3)$$

$$0 \le t \le x, \quad 0 \le y \le x, \quad 0 \le x \le x_1 \,.$$

The resolvent is a very handy function to have around.   If
it is known analytically, an integration produces the solution for any forcing function g(t).   If it is known
numerically, then a quadrature produces the same result
(assuming that the quadrature can be performed accurately).
This chapter is devoted to an initial value method for resolvents for semidegenerate kernels of the form

$$k(t,y) = \begin{cases} \displaystyle\sum_{m=1}^{M} \alpha_m(t) \, \beta_m(y) \,, & y < t \,, \\[3ex] \displaystyle\sum_{n=1}^{N} \gamma_n(t) \, \delta_n(y) \,, & t < y \,. \end{cases} \quad (3.4)$$

3.2   INITIAL VALUE PROBLEM FOR RESOLVENT

Consider the following initial value problem.   The
functions $r_{ij}(x)$ are defined by the system of ordinary
differential equations for $i=1,2,\ldots,M$; $j=1,2,\ldots,N$,

$$r_{ij}'(x) = \left[\gamma_j(x) + \sum_{m=1}^{M} \alpha_m(x) \, r_{mj}(x) \right]$$

$$\cdot \left[\beta_i(x) + \sum_{n=1}^{N} r_{in}(x) \, \delta_n(x) \right] \quad (3.5)$$

and the initial conditions

$$r_{ij}(0) = 0 . \tag{3.6}$$

We assume that the solution is bounded and continuous for $0 \leq x \leq x_1$. The functions $J_j(t,x)$, $j = 1, 2, \ldots, N$, $0 \leq t \leq x \leq x_1$, are defined by the differential equations (primes denote derivatives with respect to x)

$$J_j'(t,x) = \left[\gamma_j(x) + \sum_{m=1}^{M} \alpha_m(x)\, r_{mj}(x) \right]$$

$$\cdot \sum_{n=1}^{N} J_n(t,x)\, \delta_n(x) \tag{3.7}$$

and the initial conditions at $x = t$

$$J_j(t,t) = \gamma_j(t) + \sum_{m=1}^{M} \alpha_m(t)\, r_{mj}(t) . \tag{3.8}$$

The functions $Q_i(y,x)$, $i = 1, 2, \ldots, M$, $0 \leq y \leq x \leq x_1$, are solutions of the differential equations

$$Q_i'(y,x) = \left[\beta_i(x) + \sum_{n=1}^{N} r_{in}(x)\, \delta_n(x) \right]$$

$$\cdot \sum_{m=1}^{M} \alpha_m(x)\, Q_m(y,x) \tag{3.9}$$

with the initial conditions at $x = y$,

$$Q_i(y,y) = \beta_i(y) + \sum_{n=1}^{N} r_{in}(y) \, \delta_n(y) \; . \qquad (3.10)$$

Two auxiliary functions, Φ and Ψ, are introduced by the relations

$$\Phi(t,x) = \sum_{n=1}^{N} J_n(t,x) \, \delta_n(x) \; , \qquad (3.11)$$

$$0 \le t \le x \le x_1 \; ,$$

and

$$\Psi(y,x) = \sum_{m=1}^{M} \alpha_m(x) \, Q_m(y,x) \; , \qquad (3.12)$$

$$0 \le y \le x \le x_1 \; .$$

Finally we introduce the function K. It satisfies the differential equation

$$K'(t,y,x) = \Phi(t,x) \, \Psi(y,x) \; , \qquad (3.13)$$

$$x \ge \max(t,y) \; ,$$

and the initial conditions

$$K(t,y,t) = \Psi(y,t) \; , \qquad\qquad t \ge y, \quad (3.14)$$

$$K(t,y,y) = \Phi(t,y) \; , \qquad\qquad y > t. \quad (3.15)$$

3.3   SKETCH OF PROOF

We shall prove that the solution of the Cauchy
system is indeed the resolvent, that is, that it satisfies
the integral equation for the resolvent. The proof consists of establishing the following relations:

$$r_{ij}(x) = \int_0^x \beta_i(z) \; J_j(z,x) \; dz \; , \tag{3.16}$$

$$J_j(t,x) = \gamma_j(t) + \int_0^x k(t,z) \; J_j(z,x) \; dz, \tag{3.17}$$

$$0 \leq t \leq x \leq x_1, \quad j=1,2,\ldots,N;$$

$$r_{ij}(x) = \int_0^x Q_i(z,x) \; \gamma_j(z) \; dz \; , \tag{3.18}$$

$$i=1,2,\ldots,M; \quad j=1,2,\ldots,N; \quad 0 \leq x \leq x_1;$$

$$Q_i(y,x) = \beta_i(y) + \int_0^x Q_i(z,x) \; k(z,y) \; dz \; , \tag{3.19}$$

$$0 \leq y \leq x \leq x_1, \quad i=1,2,\ldots,M;$$

$$\Phi(t,x) = k(t,x) + \int_0^x k(t,z) \; \Phi(z,x) \; dz \; , \tag{3.20}$$

$$0 \leq t \leq x \leq x_1;$$

$$\Psi(y,x) = k(x,y) + \int_0^x \Psi(z,x) \; k(z,y) \; dz \; , \tag{3.21}$$

$$0 \leq y \leq x \leq x_1;$$

and

$$\int_0^x \beta_i(z) \ K(z,y,x) \ dz = \int_0^x Q_i(z,x) \ k(z,y) \ dz, \quad (3.22)$$

$$0 \leq y \leq x \leq x_1, \quad i=1,2,\ldots,M.$$

From these the desired relation, the integral equation (3.3), follows.

3.4   PROOF

Define the functions $s_{ij}(x)$ by the equations

$$s_{ij}(x) = \int_0^x \beta_i(z) \ J_j(z,x) \ dz \ , \qquad (3.23)$$

$$i=1,2,\ldots,M, \quad j=1,2,\ldots,N, \quad 0 \leq x \leq x_1.$$

We see that at $x = 0$ the function $s_{ij}(x)$ satisfies the relation

$$s_{ij}(0) = 0 \ . \qquad (3.24)$$

In addition, differentiation with respect to x shows that

$$s'_{ij}(x) = \beta_i(x) \ J_j(x,x) + \int_0^x \beta_i(z) \ J'_j(z,x) \ dz. \quad (3.25)$$

Using Eq. (3.7) this becomes

$$s'_{ij}(x) = \beta_i(x) \, J_j(x,x)$$

$$+ \left[\gamma_j(x) + \sum_{m=1}^{M} \alpha_m(x) \, r_{mj}(x) \right] \tag{3.26}$$

$$\cdot \sum_{n=1}^{N} \int_0^x \beta_i(z) \, J_n(z,x) \, \delta_n(x) \, dz \, .$$

In view of Eqs. (3.8) and (3.23), this becomes

$$s'_{ij}(x) = \left[\gamma_j(x) + \sum_{m=1}^{M} \alpha_m(x) \, r_{mj}(x) \right]$$

$$\tag{3.27}$$

$$\cdot \left[\beta_i(x) + \sum_{n=1}^{N} s_{in}(x) \, \delta_n(x) \right] \, .$$

Comparing the differential equations in Eqs. (3.5) and (3.27) and the initial conditions in Eqs. (3.6) and (3.24), and using standard uniqueness theorems, we see that Eq. (3.16) holds.

Next let

$$w_{ij}(x) = \int_0^x Q_i(z,x) \, \gamma_j(z) \, dz \, . \tag{3.28}$$

The initial condition at $x = 0$ is

$$w_{ij}(0) = 0 \, . \tag{3.29}$$

In addition, the derivative of $w_{ij}(x)$ is

$$w'_{ij}(x) = Q_i(x,x) \, \gamma_j(x) + \int_0^x Q'_i(z,x) \, \gamma_j(z) \, dz \, . \tag{3.30}$$

According to Eq. (3.9) the last equation may be re-written as

$$w'_{ij}(x) = Q_i(x,x) \, \gamma_j(x)$$

$$+ \left[\beta_i(x) + \sum_{n=1}^{N} r_{in}(x) \, \delta_n(x) \right] \qquad (3.31)$$

$$\cdot \sum_{m=1}^{M} \alpha_m(x) \int_0^x Q_m(z,x) \, \gamma_j(z) \, dz \; .$$

Employing Eqs. (3.10) and (3.28) this becomes

$$w'_{ij}(x) = \left[\beta_i(x) + \sum_{n=1}^{N} r_{in}(x) \, \delta_n(x) \right]$$

$$\qquad (3.32)$$

$$\cdot \left[\gamma_j(x) + \sum_{m=1}^{M} \alpha_m(x) \, w_{mj}(x) \right] \; .$$

Comparing the initial conditions in Eqs. (3.6) and (3.29) and the differential equations in Eqs. (3.5) and (3.32), we see that Eq. (3.18) is valid.

Let the function $M_j(t,x)$ be introduced as

$$M_j(t,x) = \gamma_j(t) + \int_0^x k(t,z) \, J_j(z,x) \, dz \; , \qquad (3.33)$$

$$j=1,2,\ldots,M.$$

Differentiation shows that

$$M'_j(t,x) = k(t,x) \, J_j(x,x) + \int_0^x k(t,z) \, J'_j(z,x) \, dz. \qquad (3.34)$$

From Eq. (3.7) it follows that

$$M_j^!(t,x) = k(t,x) \; J_j(x,x)$$

$$+ \left[\gamma_j(x) + \sum_{m=1}^{M} \alpha_m(x) \; r_{mj}(x) \right] \tag{3.35}$$

$$\cdot \; \sum_{n=1}^{N} \delta_n(x) \int_0^x k(t,z) \; J_n(z,x) \; dz \; .$$

Then Eq. (3.8) implies

$$M_j^!(t,x) = \left[\gamma_j(x) + \sum_{m=1}^{M} \alpha_m(x) \; r_{mj}(x) \right]$$

$$\tag{3.36}$$

$$\cdot \left[k(t,x) + \sum_{n=1}^{N} \delta_n(x) \int_0^x k(t,z) \; J_n(z,x) \; dz \right].$$

Noting the form of the kernel k in Eq. (3.4) we find

$$M_j^!(t,x) = \left[\gamma_j(x) + \sum_{m=1}^{M} \alpha_m(x) \; r_{mj}(x) \right]$$

$$\cdot \left[\sum_{n=1}^{N} \gamma_n(t) \; \delta_n(x) \right. \tag{3.37}$$

$$\left. + \sum_{n=1}^{N} \delta_n(x) \int_0^x k(t,z) \; J_n(z,x) \; dz \right] \; ,$$

or

$$M_j^!(t,x) = \left[\gamma_j(x) + \sum_{m=1}^{M} \alpha_m(x)\, r_{mj}(x) \right]$$

$$\cdot \sum_{n=1}^{N} \delta_n(x) \left[\gamma_n(t) + \int_0^X k(t,z)\, J_n(z,x)\, dz \right]. \tag{3.38}$$

Recalling the definition of the function M_j in Eq. (3.33) we see that it satisfies the differential equation

$$M_j^!(t,x) = \left[\gamma_j(x) + \sum_{m=1}^{M} \alpha_m(x)\, r_{mj}(x) \right]$$

$$\cdot \sum_{n=1}^{N} \delta_n(x)\, M_n(t,x)\ , \tag{3.39}$$

$$x \geq t, \quad j=1,2,\ldots,N.$$

The initial condition at $x = t$ is given by

$$M_j(t,t) = \gamma_j(t) + \int_0^t k(t,z)\, J_j(z,t)\, dz$$

$$= \gamma_j(t) + \int_0^t \sum_{m=1}^{M} \alpha_m(t)\, \beta_m(z)\, J_j(z,t)\, dz$$

$$= \gamma_j(t) + \sum_{m=1}^{M} \alpha_m(t) \int_0^t \beta_m(z)\, J_j(z,t)\, dz \tag{3.40}$$

$$= \gamma_j(t) + \sum_{m=1}^{M} \alpha_m(t)\, r_{mj}(t)\ .$$

The last equality follows from Eq. (3.16). Since the functions J_j and M_j satisfy the same system of differential equations and initial conditions at $x = t$, we conclude that Eq. (3.17) holds.

To prove Eq. (3.19) we introduce

$$Z_i(y,x) = \beta_i(y) + \int_0^x Q_i(z,x) \, k(z,y) \, dz \, , \qquad (3.41)$$

$$0 \leq y \leq x \leq x_1, \quad i=1,2,\ldots,M.$$

Differentiation of both sides of this equation leads to the relation

$$Z_i'(y,x) = Q_i(x,x) \, k(x,y) + \int_0^x Q_i'(z,x) \, k(z,y) \, dz. \qquad (3.42)$$

Making use of the differential equation for the function Q_i we see that

$$Z_i'(y,x) = Q_i(x,x) \, k(x,y)$$

$$+ \left[\beta_i(x) + \sum_{n=1}^N r_{in}(x) \, \delta_n(x) \right]$$

$$\qquad (3.43)$$

$$\cdot \sum_{m=1}^M \alpha_m(x) \int_0^x Q_m(z,x) \, k(z,y) \, dz \, .$$

It follows that

$$Z_i^!(y,x) = \left[\beta_i(x) + \sum_{n=1}^{N} r_{in}(x)\, \delta_n(x) \right]$$

$$(3.44)$$

$$\cdot \left[k(x,y) + \sum_{m=1}^{M} \alpha_m(x) \int_0^x Q_m(z,x)\, k(z,y)\, dz \right]$$

and

$$Z_i^!(y,x) = \left[\beta_i(x) + \sum_{n=1}^{N} r_{in}(x)\, \delta_n(x) \right]$$

$$\cdot \left[\sum_{m=1}^{M} \alpha_m(x)\, \beta_m(y) \right. \qquad (3.45)$$

$$\left. + \sum_{m=1}^{M} \alpha_m(x) \int_0^x Q_m(z,x)\, k(z,y)\, dz \right] .$$

This can be written in the form

$$Z_i^!(y,x) = \left[\beta_i(x) + \sum_{n=1}^{N} r_{in}(x)\, \delta_n(x) \right]$$

$$(3.46)$$

$$\cdot \sum_{m=1}^{M} \alpha_m(x) \left[\beta_m(y) + \int_0^x Q(z,x)\, k(z,y)\, dz \right]$$

or

$$Z_i^!(y,x) = \left[\beta_i(x) + \sum_{n=1}^{N} r_{in}(x)\, \delta_n(x) \right]$$

$$(3.47)$$

$$\cdot \sum_{m=1}^{M} \alpha_m(x)\, Z_m(y,x) ,$$

$$x \geq y.$$

Furthermore, at $x = y$ we see that

$$Z_i(y,y) = \beta_i(y) + \int_0^y Q_i(z,y) \; k(z,y) \; dz$$

$$= \beta_i(y) + \int_0^y Q_i(z,y) \sum_{n=1}^N \gamma_n(z) \; \delta_n(y) \; dz$$

$$\hspace{9cm} (3.48)$$

$$= \beta_i(y) + \sum_{n=1}^N \delta_n(y) \int_0^y Q_i(z,y) \; \gamma_n(z) \; dz$$

$$= \beta_i(y) + \sum_{n=1}^N r_{in}(x) \; \delta_n(y) \; .$$

Standard uniqueness theorems then show that

$$Z_i(y,x) = Q_i(y,x) \; , \hspace{5cm} (3.49)$$

$$i=1,2,\ldots,M, \quad x \geq y,$$

from which Eq. (3.19) follows.

To prove Eq. (3.20) we multiply both sides of Eq. (3.17) by $\delta_j(x)$ and sum. This yields

$$\sum_{j=1}^N J_j(t,x) \; \delta_j(x) = \sum_{j=1}^N \gamma_j(t) \; \delta_j(x)$$

$$\hspace{9cm} (3.50)$$

$$+ \int_0^x k(t,z) \sum_{j=1}^N J_j(z,x) \; \delta_j(x) \; dz$$

or

$$\Phi(t,x) = k(t,x) + \int_0^x k(t,y) \; \Phi(z,x) \; dz \; , \qquad (3.51)$$

$$0 \le t \le x \; ,$$

which is the desired relation.

By multiplying both sides of Eq. (3.19) by $\alpha_i(x)$ and summing we find that

$$\sum_{i=1}^M \alpha_i(x) \; Q_i(y,x) = \sum_{i=1}^M \alpha_i(x) \; \beta_i(y)$$

$$\hspace{6cm} (3.52)$$

$$+ \int_0^x \sum_{i=1}^M \alpha_i(x) \; Q_i(z,x) \; k(z,y) \; dz \; .$$

The last relation is equivalent to Eq. (3.21). Finally, we are ready to establish Eq. (3.22).

We have

$$\frac{d}{dx} \int_0^x \beta_i(z) \; K(z,y,x) \; dz$$

$$= \beta_i(x) \; K(x,y,x) + \int_0^x \beta_i(z) \; K'(z,y,x) \; dz$$

$$= \beta_i(x) \; \Psi(y,x) + \int_0^x \beta_i(z) \; \Phi(z,x) \; \Psi(y,x) \; dz$$

$$\hspace{6cm} (3.53)$$

$$= \Psi(y,x) \left[\beta_i(x) + \int_0^x \beta_i(z) \; \Phi(z,x) \; dz \right]$$

$$= \Psi(y,x) \left[\beta_i(x) + \sum_{n=1}^N r_{in}(x) \; \delta_n(x) \right]$$

and

$$\frac{d}{dx} \int_0^x Q_i(z,x) \, k(z,y) \, dz$$

$$= Q_i(x,x) \, k(x,y) + \int_0^x Q_i'(z,x) \, k(z,y) \, dz$$

$$= Q_i(x,x) \, k(x,y) + \left[\beta_i(x) + \sum_{n=1}^{N} r_{in}(x) \, \delta_n(x) \right] \tag{3.54}$$

$$\cdot \int_0^x \Psi(z,x) \, k(z,y) \, dz \; .$$

The last equation may be re-written as

$$\frac{d}{dx} \int_0^x Q_i(z,x) \, k(z,y) \, dz$$

$$= \left[\beta_i(x) + \sum_{n=1}^{N} r_{in}(x) \, \delta_n(x) \right]$$

$$\cdot \left[k(x,y) + \int_0^x \Psi(z,x) \, k(z,y) \, dz \right] \tag{3.55}$$

$$= \left[\beta_i(x) + \sum_{n=1}^{N} r_{in}(x) \, \delta_n(x) \right] \Psi(y,x) \; .$$

In view of Eqs. (3.53) and (3.55) it follows that the integrals are equal, which is the content of Eq. (3.22).

We are now ready to complete the proof that the function K defined by the Cauchy system is the Fredholm resolvent for the kernel k. We let

$$R(t,y,x) = k(t,y) + \int_0^x k(t,z) \, K(z,y,x) \, dz \; . \tag{3.56}$$

Differentiation shows that

$$R'(t,y,x) = k(t,x) \; K(x,y,x)$$

$$+\int_0^x k(t,z) \; K'(z,y,x) \; dz$$

$$= k(t,x) \; K(x,y,x)$$

$$+\int_0^x k(t,z) \; \Phi(z,x) \; \Psi(y,x) \; dz \qquad (3.57)$$

$$= \left[k(t,x) + \int_0^x k(t,z) \; \Phi(z,x) \; dz \right] \Psi(y,x)$$

$$= \Phi(t,x) \; \Psi(y,x) \; ,$$

$$x \geq \max(t,y) \; .$$

Also when $y>t$, we have the initial condition

$$R(t,y,y) = k(t,y) + \int_0^y k(t,z) \; K(z,y,y) \; dz$$

$$= k(t,y) + \int_0^y k(t,z) \; \Phi(z,y) \; dz \qquad (3.58)$$

$$= \Phi(t,y) \; .$$

It remains to investigate the initial condition when $y<t$. In this case we have

$$R(t,y,t) = k(t,y) + \int_0^t k(t,z) \; K(z,y,t) \; dz \; , \qquad (3.59)$$

$$y<t \; .$$

We must show that the right hand side of this equation is $\Psi(y,t)$, $y<t$.

Earlier we showed that

$$Q_i(y,x) = \beta_i(y) + \int_0^x Q_i(z,x) \; k(z,y) \; dz \; . \qquad (3.60)$$

According to Eq. (3.22) we have

$$Q_i(y,x) = \beta_i(y) + \int_0^x \beta_i(z) \; K(z,y,x) \; dz \; . \qquad (3.61)$$

Multiplication of both sides of the last equation by $\alpha_i(x)$ and addition yields the relation

$$\sum_{i=1}^M \alpha_i(x) \; Q_i(y,x) = \sum_{i=1}^M \alpha_i(x) \; \beta_i(y)$$

$$\qquad (3.62)$$

$$+ \int_0^x \sum_{i=1}^M \alpha_i(x) \; \beta_i(z) \; K(z,y,x) \; dz \; ,$$

or

$$\Psi(y,x) = k(x,y) + \int_0^x k(x,z) \; K(z,y,x) \; dz \; , \qquad (3.63)$$

$$y<x \; .$$

Eqs. (3.59) and (3.63) show that

$$R(t,y,t) = \Psi(y,t), \qquad\qquad (3.64)$$

$$y<t \; .$$

We have now shown that the function R satisfies the same differential equation and initial conditions as the function K. It follows that

$$K(t,y,x) = R(t,y,x), \qquad\qquad (3.65)$$

$$0 \leq t, \quad y \leq x \leq x_1 \; ,$$

which is Eq. (3.3). Our proof is now complete.

3.5   DISCUSSION

In Chapter 2, we discussed the solution of Fredholm integral equations with semidegenerate kernels via initial value problems for the unknowns and some auxiliary functions. In this chapter, we have shown how to determine the corresponding resolvents. The validation of the Cauchy system for resolvents has been presented; we leave the derivation as an exercise.

EXERCISES

1. Beginning with the Fredholm integral equation (3.1), derive the integral equation for the resolvent kernel.
2. If $K(t,y,x)$ satisfies the integral equation (3.3) and $u(t)$ is expressed in terms of K as in Eq. (3.2), then prove that $u(t)$ satisfies the integral equation (3.1).
3. Derive the initial value problem for the resolvent.
4. Describe a computational method for producing the resolvent and solution of the integral equation (3.1) for semidegenerate kernels.
5. Write a general program following the method of Exercise 4.
6. Consider the examples in Chapter 2, and solve them using the above general program.

Chapter 4

DISPLACEMENT KERNELS

4.1 INTRODUCTION

This chapter is devoted to a study of the Fredholm integral equation

$$u(t) = g(t) + \lambda(t) \int_0^c k(|t-y|)u(y) \; dy \; ,$$

$$0 \leq t \leq c,$$

under the assumptions that the kernel k has the representation

$$k(r) = \int_0^1 e^{-r/z} \; w(z) \; dz \; ,$$

$$r > 0,$$

and that the interval length c is sufficiently small that the integral equation possesses a unique solution for this length and all smaller lengths. Our aim is to show that this integral equation is equivalent to a certain Cauchy system, which is of both analytic interest and computational utility.

101

There are several interesting special results for the case of an exponential forcing term. Let us call the solution J rather than u. The complete Cauchy system involves a family of J functions and an auxiliary function, R. Use of Gaussian quadrature leads to an effective numerical method. The solution of the Cauchy system does satisfy the original integral equation, as is demonstrated. The auxiliary function R is a function of three variables. However it can be replaced by a pair of functions, X and Y, which are functions of only two variables, and are thus important in reducing the computational load.

Another pair of important functions is the b and h functions. While these are again functions of three variables, they are extremely useful, because mere algebraic operations on these functions are sufficient for the determination of J, X, Y, R and other desired functions. We derive both the algebraic formulas and the Cauchy system for b and h. In addition, it is known that these b and h functions satisfy a system of linear singular integral equations. This implies that such integral equations are also equivalent to a Cauchy system.

A general forcing function in the integral equation with displacement kernel is considered next. Here, the complete Cauchy system involves u, J, and R or X and Y. The Cauchy system is then verified.

An alternative method for determining the solution with general forcing term is to first compute the resolvent. It also satisfies a Cauchy system. Then a quadrature yields the desired solution or solutions. Still another procedure is provided by a representation formula. Other interesting results are presented.

This theory for displacement kernels is historically related to our earlier studies in radiative transfer. A physical type of derivation of the Cauchy system is presented in Chapter 9, as well as some computational results.

4.2   THE CASE OF AN EXPONENTIAL FORCING TERM

Consider the family of Fredholm integral equations

$$J(t,x,z) = \lambda(t) \; e^{-(x-t)/z}$$

$$+ \lambda(t) \int_0^x k(|t-y|) \; J(y,x,z) \; dy , \qquad (4.1)$$

$$0 \le t \le x \le x_1 , \quad 0 \le z \le 1 ,$$

where the maximum interval length x_1 is sufficiently small that Eq. (4.1) has a unique solution for all interval lengths x in the closed interval $(0,x_1)$. It is further assumed that the kernel k may be written in the form

$$k(r) = \int_0^1 e^{-r/v} \; w(v) \; dv , \qquad r>0 . \quad (4.2)$$

In this section we derive a Cauchy system for J and an auxiliary function, R .

Differentiate both sides of Eq. (4.1) with respect to x to obtain the equation

$$J_x(t,x,z) = -\frac{1}{z} \lambda(t) \, e^{-(x-t)/z}$$

$$+ \lambda(t) \, k(x-t) \, J(x,x,z) \qquad\qquad (4.3)$$

$$+ \lambda(t) \int_0^x k(|t-y|) \, J_x(y,x,z) \, dy \; .$$

This is an integral equation for the function J_x . Note the two terms of the forcing function: an exponential of the type appearing in Eq. (4.1), and the kernel itself multiplied by the constant $J(x,x,z)$.

Introduce the function Φ to be the solution of the integral equation with λ times the kernel as the forcing function,

$$\Phi(t,x) = \lambda(t) \, k(x-t) + \lambda(t) \int_0^x k(|t-y|) \, \Phi(y,x) \, dy \; ,$$
$$0 \le t \le x \le x_1 \; . \qquad (4.4)$$

From the superposition theorem for linear integral equations, it follows that the solution of Eq. (4.3), regarded as an integral equation for the function J_x , is

$$J_x(t,x,z) = -\frac{1}{z} J(t,x,z) + J(x,x,z) \, \Phi(t,x) \; , \qquad (4.5)$$
$$0 \le t \le x, \quad 0 \le z \le 1 \; .$$

Observe also that the function Φ may be expressed in terms of the function J by means of the formula

$$\Phi(t,x) = \int_0^1 J(t,x,z') \, w(z') \, dz' \; , \qquad (4.6)$$

because its forcing function in Eq. (4.4) is expressible as

$$\lambda(t) \int_0^1 e^{-(x-t)/v} w(v) \, dv \, .$$

Thus the differential equation for the function J evaluated at the fixed point t, z being a parameter, becomes

$$J_x(t,x,z) = -\frac{1}{z} \, J(t,x,z)$$

$$+ J(x,x,z) \int_0^1 J(t,x,z') \, w(z') \, dz' \, , \quad (4.7)$$

$$x \geq t, \quad 0 \leq z \leq 1.$$

This involves the family of J functions and J at the special point $t=x$.

The function $J(x,x,z)$ is now to be considered. From the integral equation (4.1) we see by setting $t=x$ that

$$J(x,x,z) = \lambda(x) \left[1 + \int_0^x k(x-y) \, J(y,x,z) \, dy \right]. \quad (4.8)$$

We wish to avoid the integration over all y which Eq. (4.8) demands. Making use of Eq. (4.2) we may rewrite Eq. (4.8) in the form

$$J(x,x,z) = \lambda(x)$$

$$(4.9)$$

$$\cdot \left[1 + \int_0^x \int_0^1 e^{-(x-y)/v} w(v) \, dv \, J(y,x,z) \, dy \right].$$

Consider first the integration of the exponential times J. Introduce the function R by means of the definition

$$R(v,z,x) = \int_0^x e^{-(x-y)/v} J(y,x,z) \, dy, \qquad (4.10)$$

$$x \geq 0, \quad 0 \leq v,z \leq 1.$$

Then Eq. (4.9) assumes the more desirable form

$$J(x,x,z) = \lambda(x) \left[1 + \int_0^1 R(v,z,x) \, w(v) \, dv \right] . \qquad (4.11)$$

Now $J(x,x,z)$ is expressed through a straightforward quadrature of R over one of its parametric arguments.

A Cauchy system for the function R will now be obtained. Differentiate both sides of Eq. (4.10) with respect to x. The result is

$$R_x(v,z,x) = J(x,x,z)$$

$$-\frac{1}{v} \int_0^x e^{-(x-y)/v} J(y,x,z) \, dy \qquad (4.12)$$

$$+ \int_0^x e^{-(x-y)/v} J_x(y,x,z) \, dy .$$

Making use of the definition of R in Eq. (4.10) and the differential equation for J in Eq. (4.5) we see that

$$R_x(v,z,x) = J(x,x,z) - \frac{1}{v} R(v,z,x)$$

$$+ \int_0^x e^{-(x-y)/v} \left[-\frac{1}{z} J(y,x,z) \right. \tag{4.13}$$

$$\left. + J(x,x,z) \, \Phi(y,x) \right] \, dy \; .$$

Again using the definition in Eq. (4.10) to evaluate the first of the integrals, we find that

$$R_x(v,z,x) = - \left(\frac{1}{v} + \frac{1}{z} \right) R(v,z,x) \tag{4.14}$$

$$+ J(x,x,z) \left[1 + \int_0^x e^{-(x-y)/v} \Phi(y,x) \, dy \right] \; .$$

The last task is to evaluate the integral in the previous equation. This we do by using the representation for the function Φ in terms of the function J given in Eq. (4.6). The result is

$$\int_0^x e^{-(x-y)/v} \, \Phi(y,x) \, dy$$

$$= \int_0^x e^{-(x-y)/v} \int_0^1 J(y,x,u') \, w(u') \, du' \, dy \tag{4.15}$$

$$= \int_0^1 R(v,u',x) \, w(u') \, du' \; ,$$

where use has again been made of the definition of the function R in Eq. (4.10). The differential equation (4.14) for the function R becomes

$$R_x(v,z,x) = -\left(\frac{1}{v} + \frac{1}{z}\right) R(v,z,x)$$

(4.16)

$$+ J(x,x,z) \left[1 + \int_0^1 R(v,u,x)\ w(u)\ du\right].$$

Finally, making use of Eq. (4.11) for the function $J(x,x,z)$, we see that the differential equation for the function R is

$$R_x(v,z,x) = -\left(\frac{1}{v} + \frac{1}{z}\right) R(v,z,x)$$

$$+ \lambda(x) \left[1 + \int_0^1 R(v',z,x)\ w(v')\ dv'\right]$$

(4.17)

$$\cdot \left[1 + \int_0^1 R(v,z',x)\ w(z')\ dz'\right],$$

$$x \geq 0, \quad 0 \leq v, z \leq 1.$$

Furthermore the initial condition on the function R at $x = 0$ follows readily from Eq. (4.10). It is

$$R(v,z,0) = 0 ,$$

(4.18)

$$0 \leq v, z \leq 1.$$

We are now in a position to state the Cauchy system for the functions R and J. Let $0 \leq t \leq x_1$. On the interval $0 \leq x \leq t$ the function R satisfies the Cauchy system

$$R_x(v,z,x) = -\left(\frac{1}{v} + \frac{1}{z}\right) R(v,z,x)$$

$$+ \lambda(x) \left[1 + \int_0^1 R(v',z,x)\ w(v')\ dv'\right] (4.19)$$

$$\cdot \left[1 + \int_0^1 R(v,z',x)\ w(z')\ dz'\right],$$

together with the initial condition

$$R(v,z,0) = 0 . \qquad (4.20)$$

At $x = t$ the initial condition on the function J is

$$J(t,t,z) = \lambda(t) \left[1 + \int_0^1 R(v',z,t) \ w(v') \ dv' \right] , \qquad (4.21)$$

and for x in the interval (t,x_1) , the function J
satisfies the differential equation

$$J_x(t,x,z) = - \frac{1}{z} \ J(t,x,z)$$

$$+ \lambda(x) \left[1 + \int_0^1 R(v',z,x) \ w(v') \ dv' \right] \qquad (4.22)$$

$$\cdot \int_0^1 J(t,x,z') \ w(z') \ dz' .$$

$$0 \le z \le 1 .$$

4.3   NUMERICAL CONSIDERATIONS

To solve the Cauchy system just described, we employ
the method of lines. The derivatives with respect to x are
retained, but the integrals on the interval $(0,1)$ are approximated by use of a Gaussian quadrature formula. We write

$$\int_0^1 f(z') \ dz' \approx \sum_{i=1}^N f(\alpha_i) \ \beta_i . \qquad (4.23)$$

The abscissas α_1, α_2, ..., α_N are the roots of the shifted Legendre polynomial $P_N(1-2\alpha)$, and the weights β_1, β_2, ..., β_N are the corresponding Christoffel numbers. Writing,

$$R_{ij}(x) = R(\alpha_i, \alpha_j, x) , \tag{4.24}$$

$$i,j=1,2,...,N, \quad x\geq 0,$$

and

$$w_i = w(\alpha_i) , \tag{4.25}$$

$$i=1,2,...,N,$$

we find the approximate system for the functions R_{ij} ,

$$R_{ij}'(x) = -\left(\alpha_i^{-1} + \alpha_j^{-1}\right) R_{ij}(x)$$

$$+ \lambda(x) \left[1 + \sum_{m=1}^{N} R_{mj}(x) \, w_m \, \beta_m\right] \tag{4.26}$$

$$\cdot \left[1 + \sum_{m=1}^{N} R_{im}(x) \, w_m \, \beta_m\right] ,$$

and

$$R_{ij}(0) = 0 , \tag{4.27}$$

$$i,j=1,2,...,N .$$

This system of N^2 ordinary differential equations is integrated from $x = 0$ to $x = t$.

Next the functions $J_1(t,x)$, $J_2(t,x)$, ..., $J_N(t,x)$ are introduced by means of the definitions

$$J_i(t,x) = J(t,x,\alpha_i), \tag{4.28}$$

$$i=1,2,\ldots,N, \quad 0 \le t \le x \le x_1 \; .$$

Approximate initial values of $J_1(t,x)$, $J_2(t,x)$,$\ldots$,$J_N(t,x)$
at $x = t$ are provided by the formulas

$$J_j(t,t) = \lambda(t) \left[1 + \sum_{m=1}^{N} R_{mj}(t) \; w_m \; \beta_m \right], \tag{4.29}$$

$$j=1,2,\ldots,N \; .$$

Then for $x \ge t$ approximate values of the functions
$J_1(t,x)$, $J_2(t,x)$,$\ldots$,$J_N(t,x)$ are obtained by adjoining the
N ordinary differential equations

$$J_j'(t,x) = -\frac{1}{\alpha_j} \, J_j(t,x)$$

$$+ \lambda(x) \left[1 + \sum_{m=1}^{N} R_{mj}(x) \; w_m \; \beta_m \right] \tag{4.30}$$

$$\cdot \sum_{n=1}^{N} J_n(t,x) \; w_n \; \beta_n \; ,$$

$$j=1,2,\ldots,N \; .$$

to the system in Eq. (4.26) and continuing the integration
until $x = x_1$. In this way approximate numerical values
are produced for the function $R(v,z,x)$ at $v,z = \alpha_1,\alpha_2,\ldots,\alpha_N$,
$0 \le x \le x_1$. The function $J(t,x,z)$ is obtained for a fixed
value of t between 0 and x_1, for x in the interval
(t,x_1), and $z = \alpha_1,\alpha_2,\ldots,\alpha_N$. Should values of J be
desired for various values of t, $0 < t_1 < t_2 \ldots < t_M < x_1$,

the scheme is this. On the interval $(0,t_1)$ the system of differential equations (4.26) for the functions $R_{ij}, i,j=1,2,\ldots,N$, is integrated numerically, subject to the initial conditions at $x = 0$ in Eq. (4.27). At $x = t_1$ the system of differential equations (4.30) for the functions $J_1(t_1,x), J_2(t_1,x),\ldots,J_N(t_1,x)$ is adjoined to the differential equations for the functions $R_{11},\ldots,R_{NN}$. The initial conditions on the functions $J_1(t_1,x), J_2(t_1,x), \ldots,J_N(t_1,x)$ at $x = t_1$ are given in Eq. (4.29) evaluated for $t = t_1$. The integration of this system of $N^2 + N$ ordinary differential equations continues until $x = t_2$. At this point N ordinary differential equations (4.30) for the functions $J_1(t_2,x), J_2(t_2,x),\ldots,J_N(t_2,x)$ are adjoined. The initial conditions on these functions are given by Eq. (4.29) evaluated for $t = t_2$. As x passes each additional point $t_3,t_4,\ldots,t_M$, N additional ordinary dif-ferential equations for which the initial conditions are known are adjoined. On the final interval, $t_M \leq x \leq x_1$, the number of ordinary differential equations being integrated is $N^2 + MN$. For $N = 7$ and $M = 10$, a standard choice, we have $N^2 + MN = 119$, a reasonable number of ordinary differential equations with known initial conditions. In this way the functions R and J are determined numerically for $0 \leq x \leq x_1$, $v,z = \alpha_1, \alpha_2,\ldots,\alpha_N$, and $t = t_1, t_2,\ldots,t_M$.

     In a later chapter, Chapter 9, we apply this techniqu to a problem in radiative transfer and demonstrate its compu-tational effectiveness.

4.4   VALIDATION OF THE CAUCHY METHOD

Earlier we have seen that the solution of the family
of Fredholm integral equations in Eq. (4.1) satisfies the
Cauchy system in Eqs. (4.19) - (4.22). We shall now show,
conversely, that the solution of the Cauchy system in Eqs.
(4.19) - (4.22) satisfies the family of Fredholm integral
equations in Eq. (4.1). Our approach hinges on the uniqueness of the solution of several Cauchy systems. The proof
is in two steps. First we shall show that the solutions R
and J of the Cauchy system, Eqs. (4.19) - (4.22), are
related by the equality

$$R(v,z,x) = \int_0^x e^{-(x-y)/v} J(y,x,z) \, dy \, , \qquad (4.31)$$

$$0 \leq v, z \leq 1, \quad 0 \leq x \, .$$

Define the function Q by the relation

$$Q(v,z,x) = \int_0^x e^{-(x-y)/v} J(y,x,z) \, dy \, , \qquad (4.32)$$

$$0 \leq v, z \leq 1, \quad 0 \leq x.$$

We will show that Q = R. Differentiate with
respect to x to obtain the equality

$$Q_x(v,z,x) = J(x,x,z) - \frac{1}{v} Q(v,z,x)$$

$$\qquad (4.33)$$

$$+ \int_0^x e^{-(x-y)/v} J_x(y,x,z) \, dy \, .$$

Using Eq. (4.22) we see that

$$Q_x(v,z,x) = J(x,x,z) - \frac{1}{v} Q(v,z,x)$$

$$+ \int_0^x e^{-(x-y)/v} \left[- \frac{1}{z} J(y,x,z) \right. \tag{4.34}$$

$$+ J(x,x,z) \int_0^1 J(y,x,z') \, w(z') \, dz' \bigg] \, dy.$$

It follows that

$$Q_x(v,z,x) = -\left(\frac{1}{v} + \frac{1}{z} \right) Q(v,z,x) + J(x,x,z) \tag{4.35}$$

$$\cdot \left[1 + \int_0^1 w(z') \, dz' \int_0^x e^{-(x-y)/v} J(y,x,z') \, d \right.$$

This equation may then be rewritten in the form

$$Q_x(v,z,x) = -\left(\frac{1}{v} + \frac{1}{z} \right) Q(v,z,x)$$

$$+ \lambda(x) \left[1 + \int_0^1 R(v',z,x) \, w(v') \, dv' \right] \tag{4.36}$$

$$\cdot \left[1 + \int_0^1 Q(v,z',x) \, w(z') \, dz' \right] ,$$

$$x \geq 0 .$$

The initial condition on the function Q at $x = 0$, in view of its definition, is

$$Q(v,z,0) = 0 . \tag{4.37}$$

Assuming that Eqs. (4.36) and (4.37) possess a unique solution for $0 \leq x \leq x_1$, and keeping Eqs. (4.19) and (4.20) in mind, we see that

$$Q(v,z,x) \equiv R(v,z,x),\qquad (4.38)$$

$$0 \leq x \leq x_1, \quad 0 \leq v,z \leq 1.$$

This establishes Eq. (4.31).

Next introduce the function M by means of the definition

$$M(t,x,z) = \lambda(t)\, e^{-(x-t)/z}$$

$$+ \lambda(t) \int_0^x k(|t-y|)\, J(y,x,z)\, dy \,, \qquad (4.39)$$

$$0 \leq t \leq x \leq x_1, \quad 0 \leq z \leq 1 \,.$$

The function k is defined by the relation

$$k(r) = \int_0^1 e^{-r/z'}\, w(z')\, dz' \,, \qquad (4.40)$$

$$r > 0 \,,$$

and, of course, the function J is determined by the Cauchy system in Eqs. (4.19) - (4.22). We wish to show that

$$J(t,x,z) = M(t,x,z) \,, \qquad (4.41)$$

$$0 \leq t \leq x \leq x_1, \quad 0 \leq z \leq 1 \,.$$

Differentiate both sides of Eq. (4.39). This yields the equation

$$M_x(t,x,z) = -\frac{1}{z} \lambda(t) \; e^{-(x-t)/z}$$

$$+ \lambda(t) \; k(x-t) \; J(x,x,z) \qquad\qquad (4.42)$$

$$+ \lambda(t) \int_0^x k(|t-y|) \; J_x(y,x,z) \; dy \; .$$

Use of the differential equation (4.22) provides

$$M_x(t,x,z) = -\frac{1}{z} \lambda(t) \; e^{-(x-t)/z}$$

$$+ \lambda(t) \; k(x-t) \; J(x,x,z)$$

$$\qquad\qquad\qquad\qquad\qquad\qquad (4.43)$$

$$+ \lambda(t) \int_0^x k(|t-y|) \left[-\frac{1}{z} J(y,x,z) \right.$$

$$\left. + J(x,x,z) \int_0^1 J(y,x,z') \; w(z') \; dz' \right] dy \; .$$

In view of the definitions in Eqs. (4.39) and (4.40), we see
that Eq. (4.42) or (4.43) becomes

$$M_x(t,x,z) = -\frac{1}{z} M(t,x,z)$$

$$\qquad\qquad\qquad\qquad\qquad\qquad (4.44)$$

$$+ J(x,x,z) \int_0^1 M(t,x,z') \; w(z') \; dz' \; ,$$

$$x \geq t.$$

At $x = t$ we observe that

$$M(t,t,z) = \lambda(t) \left[1 + \int_0^t k(t-y) \, J(y,t,z) \, dy \right]$$

$$= \lambda(t) \left[1 + \int_0^t \int_0^1 e^{-(t-y)/z'} w(z') dz' J(y,t,z) \, dy \right]$$

$$= \lambda(t) \left[1 + \int_0^1 R(z',z,t) \, w(z') \, dz' \right] , \quad \text{(4.45)}$$

in view of Eq. (4.31). Comparison of the Cauchy systems for the functions M and J for $t \leq x \leq x_1$ shows them to be identical, which establishes Eq. (4.41), i.e., the integral equation

$$J(t,x,z) = \lambda(t) \, e^{-(x-t)/z}$$

$$+ \lambda(t) \int_0^x k(|t-y|) \, J(y,x,z) \, dy , \quad \text{(4.46)}$$

$$0 \leq t \leq x \leq x_1, \quad 0 \leq z \leq 1.$$

This shows that the solution of the Cauchy system does indeed provide a solution of the family of Fredholm integral equations in Eq. (4.1).

4.5   THE X AND Y FUNCTIONS

For the important case in which Eq. (4.1) is simply

$$J(t,x,z) = e^{-(x-t)/z}$$

$$\text{(4.47)}$$

$$+ \int_0^x k(|t-y|) \, J(y,x,z) \, dy ,$$

$$0 \leq t \leq x \leq x_1, \quad 0 \leq z \leq 1,$$

an alternative treatment is possible. In place of introducing the function R, which is a function of three variables, it is possible to carry out the discussion in terms
of two functions, X and Y, each of which is a function of
only two variables. This is of particular importance from
the computational viewpoint, since it lightens the demands
on the computer's high-speed memory and also reduces the
amount of arithmetic required.

Differentiate Eq. (4.47) with respect to x to
obtain the equation

$$J_x(t,x,z) = -\frac{1}{z} e^{-(x-t)/z} + k(x-t)\ J(x,x,z)$$

$$+ \int_0^x k(|t-y|)\ J_x(y,x,z)\ dy\ .$$

(4.48)

The solution of Eq. (4.48), viewed as an integral equation
for the function J_x, is

$$J_x(t,x,z) = -\frac{1}{z} J(t,x,z) + J(x,x,z)\ \Phi(t,x),\qquad (4.49)$$

where the function Φ is introduced as a solution of the
integral equation

$$\Phi(t,x) = k(x-t) + \int_0^x k(|t-y|)\ \Phi(y,x)\ dy\ ,\qquad (4.50)$$

$$0 \le t \le x \le x_1.$$

The kernel k has the form

$$k(r) = \int_0^1 e^{-r/z'}\ w(z')\ dz'\ ,\qquad (4.51)$$

$$r > 0\ ,$$

from which it follows that

$$\Phi(t,x) = \int_0^1 J(t,x,z')\, w(z')\, dz' \, . \tag{4.52}$$

Eq. (4.50) becomes

$$J_x(t,x,z) = -\frac{1}{z}\, J(t,x,z)$$

$$\tag{4.53}$$

$$+ J(x,x,z) \int_0^1 J(t,x,z')\, w(z')\, dz' \, ,$$

$$x \geq t \, .$$

Rather than treating the function $J(x,x,z)$ by replacing t by x in Eq. (4.47), we introduce the new function X by means of the definition

$$X(x,z) = J(x,x,z), \tag{4.54}$$

$$0 \leq x \leq x_1, \quad 0 \leq z \leq 1 \, .$$

Return to Eq. (4.47) and replace t by x-t. The result is

$$J(x-t,x,z) = e^{-t/z} + \int_0^x k(|x-t-y|)\, J(y,x,z)\, dy. \tag{4.55}$$

Next introduce a new variable of integration, y', by means of the formula

$$y' = x-y \, . \tag{4.56}$$

Eq. (4.55) becomes

$$J(x-t,x,z) = e^{-t/z} + \int_0^x k(|y'-t|) \, J(x-y',x,z) \, dy' .$$

$$(4.57)$$

Differentiate both sides of this equation with respect to x. It becomes

$$\frac{d}{dx} J(x-t,x,z) = k(x-t) \, J(0,x,z)$$

$$(4.58)$$

$$+ \int_0^x k(|y'-t|) \, \frac{d}{dx} J(x-y',x,z) \, dy'.$$

Eq. (4.58) may now be viewed as a Fredholm integral equation whose solution is

$$\frac{d}{dx} J(x-t,x,z) = J(0,x,z) \, \Phi(t,x) ,$$

$$(4.59)$$

$$0 \leq t \leq x \leq x_1.$$

Upon replacing t by 0 it becomes

$$\frac{d}{dx} J(x,x,z) = J(0,x,z) \, \Phi(0,x) .$$

$$(4.60)$$

The last equation may be rewritten in the form

$$X_x(x,z) = J(0,x,z) \int_0^1 J(0,x,z') \, w(z') \, dz' .$$

$$(4.61)$$

Introduce the function Y to be

$$Y(x,z) = J(0,x,z) \; , \tag{4.62}$$

$$0 \leq x \leq x_1, \quad 0 \leq z \leq 1 \; .$$

The differential equation for the function Y follows immediately from Eq. (4.53) by replacing t by 0. The result is

$$Y_x(x,z) = -\frac{1}{z} Y(x,z) + X(x,z) \int_0^1 Y(x,z') \, w(z') \, dz'. \tag{4.63}$$

From the integral equation (4.47) we see directly that the initial conditions are

$$X(0,z) = 1 \; , \tag{4.64}$$

$$Y(0,z) = 1 \; , \tag{4.65}$$

$$0 \leq z \leq 1.$$

It follows that the functions X and Y, defined by the Eqs. (4.54) and (4.62) in terms of the function J, satisfy the Cauchy system

$$X_x(x,z) = Y(x,z) \int_0^1 Y(x,z') \, w(z') \, dz' \; , \tag{4.66}$$

$$Y_x(x,z) = -\frac{1}{z} Y(x,z) \tag{4.67}$$

$$+ X(x,z) \int_0^1 Y(x,z') \, w(z') \, dz',$$

$$x \geq 0, \quad 0 \leq z \leq 1 \; ;$$

$$X(0,z) = 1 \; , \qquad\qquad\qquad (4.68)$$

$$Y(0,z) = 1, \qquad\qquad\qquad (4.69)$$

$$0 \leq z \leq 1 \; .$$

It follows that the function J itself, for a fixed value of t in the interval $(0,x_1)$ may be determined thusly. The Cauchy system, Eqs. (4.66) - (4.69), is integrated from $x = 0$ to $x = t$. At this point we set

$$J(t,t,z) = X(t,z) \; , \qquad\qquad (4.70)$$

$$0 \leq z \leq 1 \; .$$

The differential equation

$$J_x(t,x,z) = -\frac{1}{z} J(t,x,z)$$

$$\qquad\qquad\qquad\qquad\qquad\qquad\qquad (4.71)$$

$$+ X(x,z) \int_0^1 J(t,x,z') \; w(x') \; dz'$$

is then adjoined to the system above for the functions X and Y and the integration continues until $x = x_1$. In this way no need for the production of the function R arises.

The function X can be expressed in terms of the function R. According to its definition, for the function X we may write

$$X(x,z) = J(x,x,z)$$

$$= 1 + \int_0^x k(x-y) \; J(y,x,z) \; dy$$

(4.72)

$$= 1 + \int_0^x \int_0^1 e^{-(x-y)/z'} \; w(z') \; dz' \; J(y,x,z) \; dy$$

$$= 1 + \int_0^1 R(z',z,x) \; w(z') \; dz' \; .$$

It is also possible to express the function R in terms of the functions X and Y. In this case the relation proves to be algebraic.

First let us show that the function R, in spite of the asymmetrical appearance of the variables v and z in its definition,

$$R(v,z,x) = \int_0^x e^{-(x-y)/v} \; J(y,x,z) \; dy \; ,$$

(4.73)

is actually symmetrical in its first two arguments. We write

$$J(t,x,z) = e^{-(x-t)/z} + \int_0^x k(|t-y|) \; J(y,x,z) \; dy,$$

(4.74)

$$J(t,x,v) = e^{-(x-t)/v} + \int_0^x k(|t-y|) \; J(y,x,v) \; dy \; .$$

(4.75)

Cross-multiplication, integration on the variable t from 0 to x, and cancellation of like terms then shows that

$$\int_0^x J(t,x,z)\ e^{-(x-t)/v}\ dt = \int_0^x J(t,x,v)\ e^{-(x-t)/z}\ dt \ .$$

$$(4.76)$$

This expresses the symmetry of the function R,

$$R(v,z,x) = R(z,v,x) \ , \tag{4.77}$$

$$0 \le z, v \le 1, \quad x \ge 0.$$

The differential equation for the function R may be written in the form

$$R_x(v,z,x) = -\left(\frac{1}{v} + \frac{1}{z}\right)\ R(v,z,x)$$

$$(4.78)$$

$$+ \ X(x,z)\ X(x,v) \ .$$

A second expression for the function R_x may also be obtained. In the integral that defines the function R introduce a new variable of integration y' ,

$$y' = x-y \ . \tag{4.79}$$

The formula becomes

$$R(v,z,x) = \int_0^x e^{-y'/v}\ J(x-y',x,z)\ dy',$$

$$(4.80)$$

$$= \int_0^x e^{-y/v}\ J(x-y,x,z)\ dy \ .$$

Differentiation then shows that

$$R_x(v,z,x) = e^{-x/v} \, Y(x,z)$$

$$+ \int_0^x e^{-y/v} \frac{d}{dx} \, J(x-y,x,z) \, dy \tag{4.81}$$

or

$$R_x(v,z,x) = Y(x,z) \left[e^{-x/v} + \int_0^x e^{-y/v} \, \Phi(y,x) \, dy \right]. \tag{4.82}$$

To transform the integral in the last equation we cross-multiply Eqs. (4.54) and (4.57), integrate, and cancel like terms to obtain the equation

$$\int_0^x e^{-y/v} \, \Phi(y,x) \, dy = \int_0^x J(x-y,x,v) \, k(x-y) \, dy \, . \tag{4.83}$$

It follows that

$$R_x(v,z,x) = Y(x,z) \left[e^{-x/v} + \int_0^x J(x-y,x,v) \, k(x-y) \, dy \right] \tag{4.84}$$

$$= Y(x,z) \, Y(x,v) \, .$$

Equating the two expressions obtained for the function R_x yields the result

$$-\left(\frac{1}{v} + \frac{1}{z} \right) R(v,z,x) + X(x,z) \, X(x,v) = Y(x,z) \, Y(x,v) \, , \tag{4.85}$$

$$R(v,z,x) = \frac{X(x,z) \, X(x,v) - Y(x,z) \, Y(x,v)}{v^{-1} + z^{-1}} \, . \tag{4.86}$$

This formula explicitly exhibits the symmetry of the function R.

4.6   THE FUNCTIONS b AND h

Much of our attention has been devoted to the integral equation

$$J(t,x,z) = e^{-(x-t)/z} + \int_0^x k(|t-y|) \, J(y,x,z) \, dy \, ,$$

$$\text{(4.87)}$$

$$0 \leq t \leq x, \quad 0 \leq z \leq 1.$$

It is well to observe, though, that frequently our interest lies not so much in the solution function J itself, as in certain functionals on J. We have already seen the important roles played by

$$X(x,z) = J(x,x,z) \, , \tag{4.88}$$

$$Y(x,z) = J(0,x,z) \, , \tag{4.89}$$

$$R(v,z,x) = \int_0^x e^{-(x-y)/v} \, J(y,x,z) \, dy \, , \tag{4.90}$$

$$\Phi(t,x) = \int_0^1 J(t,x,z') \, w(z') \, dz' \, . \tag{4.91}$$

Of equally great significance is the function I, defined by the equation

$$I(t,v,x,u) = \begin{cases} v^{-1} \int_0^t e^{-(t-y)/v} \, J(y,x,u) \, dy, & v>0 \, , \\[2ex] -v^{-1} \int_t^x e^{-(t-y)/v} \, J(y,x,u) \, dy, & v<0 \, , \end{cases}$$

$$\text{(4.92)}$$

$$0 \leq t \leq x, \quad 0 \leq u \leq 1 \, .$$

This function plays an important role in radiative transfer and in filtering.

In this section two new functions, b and h, are introduced. It will be shown that the functions mentioned above, as well as the function J itself, may all be expressed *algebraically* in terms of b and h. These functions depend on just three variables, so they are easier to deal with computationally than the function I. In addition it will be seen that these two are actually one, and that they satisfy a Cauchy system which allows them to be readily obtained numerically.

The functions b and h are defined in terms of the function Φ by the relations

$$b(t,v,x) = \begin{cases} v^{-1} \int_0^t e^{-(t-y)/v} \, \Phi(y,x) \, dy, & v>0 \, , \\[2ex] -v^{-1} \left[e^{-(t-x)/v} + \int_t^x e^{-(t-y)/v} \, \Phi(y,x) \, dy \right], & v<0 \, , \end{cases}$$

$$\text{(4.93)}$$

$$h(t,v,x) = \begin{cases} v^{-1} \left[e^{-t/v} + \int_0^t e^{-(t-y)/v} \, \Phi(x-y,x) \, dy \right], & v>0 \, , \\[2ex] -v^{-1} \int_t^x e^{-(t-y)/v} \, \Phi(x-y,x) \, dy, & v<0 \, , \end{cases}$$

$$0 \leq t \leq x, \quad -1 \leq v \leq +1.$$

$$\text{(4.94)}$$

As usual, the function Φ is defined by the integral equation

$$\Phi(t,x) = k(x-t) + \int_0^x k(|t-y|) \; \Phi(y,x) \; dy \; , \qquad (4.95)$$

$$0 \le t \le x \le x_1 \; .$$

The representations to be shown are:

$$X(x,u) = 1 + ub(x,u,x) \; , \qquad (4.96)$$

$$Y(x,u) = uh(x,u,x) \; , \qquad (4.97)$$

$$\left(\frac{u+v}{u}\right) I(t,v,x,u) = [1 + ub(x,u,x)]$$

$$\cdot [ub(t,-u,x) + vb(t,v,x)] \qquad (4.98)$$

$$- uh(x,u,x) \; [uh(t,-u,x) + vh(t,v,x)],$$

$$J(t,x,u) = u \{ [1 + ub(x,u,x)] \; b(t,-u,x)$$

$$\qquad (4.99)$$

$$- uh(x,u,x) \; h(t,-u,x) \} \; ,$$

$$\Phi(t,x) = b(t,0,x) \; . \qquad (4.100)$$

For the functions r and τ, defined by the equations,

$$r(v,u,x) = I(x,v,x,u), \qquad (4.101)$$

$$v > 0 \; ,$$

$$\tau(v,u,x) = I(0,-v,x,u), \qquad (4.102)$$

$$v > 0 \; ,$$

we have the additional representations

$$\left(\frac{u+v}{u}\right) r(v,u,x) = [1 + ub(x,u,x)] [1 + vb(x,v,x)]$$

$$- uh(x,u,x) vh(x,v,x) , \qquad (4.103)$$

and

$$\left(\frac{u-v}{u}\right) \tau(v,u,x) = uh(x,u,x) [1 + vb(x,v,x)]$$

$$\qquad (4.104)$$

$$- [1 + ub(x,u,x)] vh(x,v,x) .$$

In addition, the functions b and h are inter-connected by the functional relationships

$$b(t, + v,x) = h(x-t,-v,x) , \qquad (4.105)$$

$$h(t, + v,x) = b(x-t,-v,x) , \qquad (4.106)$$

$$0 \leq t \leq x, \quad -1 \leq v \leq +1 .$$

They enable us to confine our attention to the case in which the second argument of the functions b and h is negative, an important point computationally, as we shall see later.

The Cauchy system for the functions b and h will be seen to be

$$b_x(t,v,x) = \Phi(0,x) h(t,v,x)$$

$$\qquad (4.107)$$

$$+ v^{-1} \{b(t,v,x) - \Phi(t,x)\}$$

$$h_x(t,v,x) = \Phi(0,x) b(t,v,x) , \qquad (4.108)$$

$$x \geq t .$$

The initial conditions at $x = t$ are

$$b(t,v,t) = -v^{-1} , \tag{4.109}$$

$$h(t,v,t) = 0 , \tag{4.110}$$

$$v<0 .$$

4.7   A DECOMPOSITION FORMULA

The first goal in the derivation is to obtain an expression for the function J_t . Write Eq. (4.87) in the form

$$J(t,x,u) = e^{-(x-t)/u} + \int_0^t k(t-y) \; J(y,x,u) \; dy$$
$$+ \int_t^x k(y-t) \; J(y,x,u) \; dy , \tag{4.111}$$

and differentiate with respect to t. This results in the equation

$$J_t(t,x,u) = u^{-1} \, e^{-(x-t)/u} + \int_0^t k'(t-y) \; J(y,x,u) \; dy$$
$$- \int_t^x k'(y-t) \; J(y,x,u) \; dy . \tag{4.112}$$

As usual, the prime denotes differentiation of the function k with respect to its single argument. Through integration by parts the last equation becomes

$$J_t(t,x,u) = u^{-1} e^{-(x-t)/u} - J(y,x,u) \; k(t-y) \; \Big|_0^t$$

$$+ \int_0^t k(t-y) \; J_t(y,x,u) \; dy$$

$$- J(y,x,u) \; k(y-t) \; \Big|_0^t \qquad\qquad (4.113)$$

$$+ \int_t^x k(y-t) \; J_t(y,x,u) \; dy \; ,$$

or

$$J_t(t,x,u) = u^{-1} e^{-(x-t)/u} + J(0,x,u) \; k(t)$$

$$\qquad\qquad\qquad\qquad\qquad\qquad\qquad\qquad (4.114)$$

$$- J(x,x,u) \; k(x-t) + \int_0^x k(|t-y|) \; J_t(y,x,u) \; dy \; .$$

Regard this as an integral equation for the function J_t.
Its solution is

$$J_t(t,x,u) = u^{-1} J(t,x,u) - X(x,u) \; \Phi(t,x)$$

$$+ Y(x,u) \; \Phi(x-t,x), \qquad\qquad (4.115)$$

$$0 \leq t \leq x, \quad 0 \leq u \leq 1 \; .$$

We may now use the expression for the function J_t
in Eq. (4.115) and the definitions of the functions I, b   and
h   in Eqs. (4.92) - (4.94), together with integration by parts,
to obtain the representation in Eq. (4.98) for the function
I   in terms of b   and h .   First consider the case in which
v   is non-negative,

$$v I(t,v,x,u) = \int_0^t e^{-(t-y)/v} \; J(y,x,u) \; dy \; , \qquad (4.116)$$

$$0 \leq t \leq x.$$

It follows that

$$vI(t,v,x,u) = J(y,x,u) \; v \; e^{-(t-y)/v} \Big|_0^t$$

(4.117)

$$- v \int_0^t e^{-(t-y)/v} \; J_t(y,x,u) \; dy \; .$$

From this we see that

$$vI(t,v,x,u) = v \; J(t,x,u) \; - \; vJ(0,x,u) \; e^{-t/v}$$

$$- v \int_0^t e^{-(t-y)/v}$$

(4.118)

$$\left\{ u^{-1} \; J(y,x,u) \; - \; J(x,x,u) \; \Phi(y,x) \right.$$

$$\left. + \; J(0,x,u) \; \Phi(x-y,x) \right\} \; dy \; .$$

This becomes

$$I(t,v,x,u) = J(t,x,u) \; - \; \frac{v}{u} I(t,v,x,u)$$

$$+ \; v \; X(x,u) \; b(t,v,x)$$

(4.119)

$$- \; Y(x,u) \left[e^{-t/v} + \int_0^t e^{-(t-y)/v} \; \Phi(x-y,x) \, dy \right],$$

$$\left(1 + \frac{v}{u} \right) I(t,v,x,u) = J(t,x,u) \; + \; vX(x,u) \; b(t,v,x)$$

(4.120)

$$- \; Y(x,u) \; vh(t,v,x) \; ,$$

$$v>0 \; .$$

Observe that the function I has now been expressed
in terms of the functions J, X, Y, b and h for v>0. The

same formula holds when $v<0$. To see this write

$$-vI(t,v,x,u) = \int_t^x e^{-(t-y)/v} J(y,x,u) \; dy$$

$$= v \; J(y,x,u) \; e^{-(t-y)/v} \Big|_t^x$$

$$- v \int_t^x e^{-(t-y)/v} \; J_t(y,x,u) \; dy$$

$$= J(x,x,u) \; ve^{-t-x)/v} \; -vJ(t,x,u) \tag{4.121}$$

$$- v \int_t^x e^{-(t-y)/v}$$

$$\left\{ u^{-1} \; J(y,x,u) \; - \; X(x,u) \; \Phi(y,x) \right.$$

$$\left. + \; Y(x,u) \; \Phi(x-y,x) \right\} dy \; ,$$

$$v<0 \; .$$

It is then seen that

$$-vI(t,v,x,u) = v \; X(x,u) \; e^{-(t-x)/v} \; -v \; J(t,x,u)$$

$$- v \; \frac{-v}{u} \; I(t,v,x,u)$$

$$\tag{4.122}$$

$$+ \; v \; X(x,u) \int_t^x e^{-(t-y)/v} \; \Phi(y,x) \; dy$$

$$- \; v \; Y(x,u) \int_t^x e^{-(t-y)/v} \; \Phi(x-y,x) \; dy \; ,$$

$$\frac{u + v}{u} I(t,v,x,u) = J(t,x,u)$$

$$- X(x,u) \left\{ e^{-(t-x)/v} + \int_t^x e^{-(t-y)/v} \Phi(y,x) \, dy \right\}$$

(4.123)

$$+ Y(x,u) \int_t^x e^{-(t-y)/v} \Phi(x-y,x) \, dy \ ,$$

$$\frac{u + v}{u} I(t,v,x,u) = J(t,x,u) + vb(t,v,x) \, X(x,u)$$

$$- v \, Y(x,u) \, h(t,v,x) \ ,$$

(4.124)

$$v<0 \ .$$

We have now seen that

$$\left(1 + \frac{v}{u}\right) I(t,v,x,u) = J(t,x,u) + X(x,u) \, vb(t,v,x)$$

$$-Y(x,u) \, vh(t,v,x) \ ,$$

(4.125)

$$0 \le u \le 1, \quad -1 \le v \le +1, \quad 0 \le t \le x \ .$$

By putting $v = -u$ in the above equation we find that

$$J(t,x,u) = ub(t,-u,x) \, X(x,u)$$

$$- uh(t,-u,x) \, Y(x,u) \ ,$$

(4.126)

$$0 \le t \le x, \quad 0 \le u \le 1 \ .$$

This expresses J in terms of X, Y, b and h. It remains but to express the functions X and Y in terms of the functions b and h.

4.8   X   AND   Y   IN TERMS OF   b   AND   h, AND RELATED MATTERS

Recall that

$$X(x,u) = J(x,x,u) \qquad (4.127)$$

and

$$Y(x,u) = J(0,x,u) \qquad (4.128)$$

From the integral equation for the function J we see that

$$X(x,u) = 1 + \int_0^x k(x-y) \; J(y,x,u) \; dy \qquad (4.129)$$

and

$$Y(x,u) = e^{-x/u} + \int_0^x k(y) \; J(y,x,u) \; dy \; . \qquad (4.130)$$

Keeping in mind the integral equations for the functions   J
and Φ and employing cross multiplication lead to the
formulas

$$X(x,u) = 1 + \int_0^x \Phi(y,x) \; e^{-(x-y)/u} \; dy \qquad (4.131)$$

and

$$Y(x,u) = e^{-x/u} + \int_0^x \Phi(x-y,x) \; e^{-(x-y)/u} \; dy \; , \qquad (4.132)$$

$$u>0 \; .$$

Then from the definitions of the functions   b   and   h   it is
seen that

$$X(x,u) = 1 + ub(x,u,x) \tag{4.133}$$

and

$$Y(x,u) = uh(x,u,x) \ , \tag{4.134}$$

which are the desired representations for the functions X and Y in terms of the functions b and h. Eqs. (4.96) and (4.97) have now been established. To obtain Eq. (4.99) we have but to substitute these results in Eq. (4.126). Similarly, the desired representation for the function I in Eq. (4.98) is obtained from Eq. (4.120) and the above results. Eqs. (4.103) and (4.104) for the functions r and τ , defined in terms of the function I, follow immediately.

By a direct substitution and comparison we see that

$$I(t,0,x,u) = J(t,x,u), \tag{4.135}$$

$$0 \le t \le x, \quad 0 \le u \le 1 \ .$$

Also note, from their definitions, that

$$b(t,v,x) = \int_0^1 I(t,v,x,u) \ w(u) \ du \ , \tag{4.136}$$

$$v > 0.$$

Taking the limit of both sides of this equation as v tends to zero through positive values yields the equality

$$b(t,0,x) = \lim_{v \to 0} \int_0^1 I(t,v,x,u) \ w(u) \ du \ . \tag{4.137}$$

Assuming that we may interchange the orders of integration
and passing to the limit yields

$$b(t,0,x) = \int_0^1 I(t,0,x,u) \; w(u) \; du$$

(4.138)

$$= \int_0^1 J(t,x,u) \; w(u) \; du \; .$$

This leads to the desired representation formula

$$\Phi(t,x) = b(t,0,x) \; ,$$

(4.139)

$$0 \leq t \leq x \; .$$

Next let us show that the functions b and h are
actually one. They are interrelated by the equations

$$b(t,-v,x) = h(x-t,v,x)$$

(4.140)

or

$$h(t,-v,x) = b(x-t,v,x) \; ,$$

(4.141)

$$0 \leq t \leq x, \quad -1 \leq v \leq +1 \; .$$

These functional relationships enable us to confine our attention, in the calculation of the functions b and h, to
values of the second argument that are negative. These
relations are established directly from the definitions of
the functions b and h. For example, for $v>0$ we have

$$b(t,-v,x) = v^{-1} \left\{ e^{(t-x)/v} + \int_t^x e^{(t-y)/v} \, \Phi(y,x) \, dy \right\}$$

$$= v^{-1} \left\{ e^{-(x-t)/v} + \int_0^{x-t} e^{-(x-t-y)/v} \, \Phi(x-y,x) \, d \right.$$

(4.142)

On the other hand we have, for $v>0$

$$h(x-t,v,x) = v^{-1} \left\{ e^{-(x-t)/v} + \right.$$

(4.143)

$$\left. + \int_0^{x-t} e^{-(x-t-y)/v} \, \Phi(x-y,x) \, dy \right\}.$$

For $v<0$ a similar validation is possible. This establishes
the functional relationship in Eq. (4.140), and Eq. (4.141)
is a simple restatement.

4.9   A CAUCHY SYSTEM FOR   b   AND   h

We now derive the Cauchy system for the functions   b
and   h.   Consider the formula

$$h(t,v,x) = -v^{-1} \int_t^x e^{-(t-y)/v} \, \Phi(x-y,x) \, dy \, , \qquad (4.144)$$

$$v<0 \, .$$

Keep in mind that

$$\frac{d}{dx} J(x-t,x,z) = Y(x,z) \, \Phi(t,x) \, , \qquad (4.145)$$

so that

$$\frac{d}{dx} \Phi(x-t,x) = \Phi(0,x) \Phi(t,x) .$$

(4.146)

Through differentiation with respect to x Eq. (4.144) becomes

$$h_x(t,v,x) = -v^{-1} \left\{ e^{-(t-x)/v} \, \phi(0,x) \right.$$

$$\left. + \int_t^x e^{-(t-y)/v} \, \phi(0,x) \, \phi(y,x) \, dy \right\},$$

(4.147)

or

$$h_x(t,v,x) = \Phi(0,x) \, b(t,v,x) ,$$

(4.148)

$$v<0 .$$

The same relation holds for $v>0$.

Next consider the relation for b when $v>0$,

$$b(t,v,x) = v^{-1} \int_0^t e^{-(t-y)/v} \, \Phi(y,x) \, dy .$$

(4.149)

To take advantage of Eq. (4.146) we make the substitution

$$y = x - y' ,$$

(4.150)

so that

$$b(t,v,x) = v^{-1} \int_{x-t}^x e^{-(t-x+y')/v} \, \Phi(x-y',x) \, dy' .$$

(4.151)

Then differentiation yields

$$b_x(t,v,x) = v^{-1} \left\{ e^{-t/v} \, \Phi(0,x) - \Phi(t,x) \right\}$$

$$+ v^{-1} \left\{ v^{-1} \int_{x-t}^{x} e^{-(t-x+y')/v} \, \Phi(x-y',x) \, dy' \right.$$

$$+ \int_{x-t}^{x} e^{-(t-x+y')/v} \, \Phi(0,x) \, \Phi(y',x) \, dy' \left. \right\} \, .$$

$$(4.152)$$

This becomes

$$b_x(t,v,x) = -v^{-1} \, \Phi(t,x) + v^{-1} \, b(t,v,x)$$

$$(4.153)$$

$$+ \Phi(0,x) \, h(t,v,x) \, , \qquad\qquad v>0 \, .$$

The same differential equation holds when $v<0$.

The initial conditions in Eqs. (4.109) and (4.110) follow directly from setting $x = t$ in the definitions of b and h in Eqs. (4.93) and (4.94).

4.10   INTEGRAL EQUATIONS FOR b AND h

Integral equations for the functions X, Y, b and h are readily obtained from the formulas derived in this section. Consider first the functions X and Y. We have

$$X(x,v) = 1 + vb(x,v,x) \, , \qquad\qquad v>0. \quad (4.154)$$

This may be rewritten in the form

$$X(x,v) = 1 + v \int_0^1 I(x,v,x,u) \; w(u) \; du$$

$$(4.155)$$

$$= 1 + v \int_0^1 r(v,u,x) \; w(u) \; du \; ,$$

or, finally,

$$X(x,v) = 1$$

$$(4.156)$$

$$+ v \int_0^1 [X(x,u) \; X(x,v) - Y(x,u) \; Y(x,v)] \; \frac{uw(u)}{u+v} \; du \; .$$

For the function Y we write

$$Y(x,v) = vh(x,v,x)$$

$$= v \left\{ \int_0^1 I(0,-v,x,u) \; w(u) \; du + \frac{1}{v} e^{-x/v} \right\},$$

$$(4.157)$$

$$= e^{-x/v} + v \int_0^1 \tau(v,u,x) \; w(u) \; du,$$

$$v > 0 \; .$$

The final result is that

$$Y(x,v) = e^{-x/v}$$

$$(4.158)$$

$$+ v \int_0^1 [Y(x,u) \; X(x,v) - X(x,u) \; Y(x,v)] \; \frac{uw(u)}{u-v} \; du \; .$$

The Eqs. (4.156) and (4.158) form a nonlinear system of integral equations for the functions X and Y.

In this paragraph let v be such that

$$0 \le v \le 1 \ . \tag{4.159}$$

Observe that

$$b(t,v,x) = \int_0^1 I(t,v,x,u) \ w(u) \ du \ . \tag{4.160}$$

Also note that the representation formula (4.98) for the function I may be written in the form

$$
I(t,v,x,u) = \frac{u}{u+v} [\ u \ X(x,u) \ b(t,-u,x)
$$
$$
- \ u \ Y(x,u) \ h(t,-u,x) + v \ X(x,u) \ b(t,v,x)
$$
$$
- \ v \ Y(x,u) \ h(t,v,x) \ ] \quad . \tag{4.161}
$$

It follows that

$$
b(t,v,x) = \int_0^1 [u \ X(x,u) \ b(t,-u,x)
$$
$$
- \ uY(x,u) \ h(t,-u,x) + vX(x,u) \ b(t,v,x)
$$
$$
- \ vY(x,u) \ h(t,v,x)] \ \frac{uw(u)}{u+v} \ du \ . \tag{4.162}
$$

Also we see that

$$b(t,-v,x) = \frac{1}{v} e^{-(x-t)/v} + \int_0^1 I(t,-v,x,u) \; w(u) \; du$$

$$= v^{-1} e^{-(x-t)/v} + \int_0^1 \Big\{ u \; X(x,u) \; b(t,-u,x)$$

$$\text{(4.163)}$$

$$-uY(x,u) \; h(t,-u,x) + vY(x,u) \; h(t,-v,x)$$

$$-vX(x,u) \; b(t,-v,x) \Big\} \; \frac{uw(u)}{u-v} \; du \; .$$

For the function $h(t,v,x)$ we have

$$h(t,v,x) = v^{-1}e^{-t/v} + \int_0^1 I(x-t,-v,x,u) \; w(u) \; du \; ,$$

$$\text{(4.164)}$$

$$h(t,v,x) = v^{-1}e^{-t/v} + \int_0^1 \Big\{ uX(x,u) \; h(t,u,x)$$

$$\text{(4.165)}$$

$$+ vY(x,u) \; b(t,v,x) - vX(x,u) \; h(t,v,x)$$

$$- uY(x,u) \; b(t,u,x) \Big\} \; \frac{uw(u)}{u-v} \; du \; .$$

And finally we note that

$$h(t,-v,x) = \int_0^1 I(x-t,v,x,u) \; w(u) \; du \; , \qquad \text{(4.166)}$$

$$h(t,-v,x) = \int_0^1 \left\{ uX(x,u)\ h(t,u,x) \right.$$

$$(4.167)$$

$$- uY(x,u)\ b(t,u,x) + vX(x,u)\ h(t,-v,x)$$

$$\left. - vY(x,u)\ b(t,-v,x) \right\} \frac{uw(u)}{u+v}\ du \ .$$

Assuming that the functions X and Y are known and that t and x are fixed with $0 \leq t \leq x$ and $0 \leq v \leq 1$, the Eqs. (4.16 (4.163), (4.165) and (4.167) form a linear system of integral equations for the functions $b(t,v,x)$, $b(t,-v,x)$, $h(t,v,x)$, and $h(t,-v,x)$.

4.11   A GENERAL FORCING FUNCTION

We now focus our attention on the family of Fredholm integral equations

$$u(t,x) = g(t) + \lambda(t) \int_0^x k(|t-y|)\ u(y,x)\ dy \ , \qquad (4.168)$$

$$0 \leq t \leq x \leq x_1 \ ,$$

where, as earlier, the kernel is given by Eq. (4.2). The first task is to find a Cauchy system satisfied by the function u. Differentiate both sides of Eq. (4.168) to obtain the relation

$$u_x(t,x) = \lambda(t)\ k(x-t)\ u(x,x)$$

$$(4.169)$$

$$+ \lambda(t) \int_0^x k(|t-y|)\ u_x(y,x)\ dy \ .$$

Regarding this as an integral equation for the function u_x, and keeping Eq. (4.4) for the function Φ in mind, we see that

$$u_x(t,x) = \Phi(t,x)\, u(x,x),\qquad\qquad (4.170)$$

$$x \geq t\ .$$

In our earlier discussion we produced an initial value problem for the function J and expressed the function Φ in terms of the function J (Eq. (4.6)). It follows that we may confine ourselves to a discussion of the second factor on the right hand side of the previous equation, $u(x,x)$. According to Eq. (4.168) we may write

$$u(x,x) = g(x) + \lambda(x) \int_0^x k(x-y)\, u(y,x)\, dy\ .\qquad (4.171)$$

Using Eq. (4.2) for the function k to transform this last equation into the form

$$u(x,x) = g(x)\qquad\qquad\qquad\qquad\qquad\qquad (4.172)$$

$$+ \lambda(x) \int_0^x \int_0^1 e^{-(x-y)/z'}\, w(z')\, dz'\, u(y,x)\, dy\ .$$

Then introduce a new function e by means of the definition

$$e(z,x) = \int_0^x e^{-(x-y)/z}\, u(y,x)\, dy,\qquad\qquad (4.173)$$

$$0 \leq x \leq x_1,\quad 0 \leq z \leq 1\ ,$$

so that Eq. (4.172) becomes

$$u(x,x) = g(x) + \lambda(x) \int_0^1 e(z',x) \, w(z') \, dz' . \qquad (4.174)$$

We shall now obtain a differential equation for the function e . Differentiation of both sides of Eq. (4.173) shows that

$$e_x(z,x) = -\frac{1}{z} e(z,x) + u(x,x)$$

$$+ \int_0^x e^{-(x-y)/z} u_x(y,x) \, dy . \qquad (4.175)$$

In view of the differential equation (4.170) this equation becomes

$$e_x(z,x) = -\frac{1}{z} e(z,x) + u(x,x)$$

$$\cdot \left[1 + \int_0^x e^{-(x-y)/z} \Phi(y,x) \, dy \right] . \qquad (4.176)$$

Earlier, in Eq. (4.15), we found that

$$\int_0^x e^{-(x-y)/z} \Phi(y,x) \, dy = \int_0^1 R(z,u',x) \, w(u') \, du' . \qquad (4.177)$$

It follows that the function e satisfies the differential equation

$$e_x(z,x) = -\frac{1}{z} e(z,x)$$

$$+ \left[g(x) + \lambda(x) \int_0^1 e(z',x) \; w(z') \; dz' \right] \qquad (4.178)$$

$$\cdot \left[1 + \int_0^1 R(z,u',x) \; w(u') \; du' \right] \; .$$

From the definition of the function e in Eq. (4.173) it is seen that the initial condition at $x = 0$ is

$$e(z,0) = 0, \qquad\qquad\qquad\qquad (4.179)$$
$$0 \leq z \leq 1 \; .$$

All of the required differential equations and initial conditions for determining the function u have now been obtained. The Cauchy system may now be stated.

Let t be a fixed value for which $0 \leq t \leq x_1$. On the interval $(0,t)$ the functions R and e are determined as solutions of the Cauchy system

$$R_x(v,z,x) = -\left(\frac{1}{v} + \frac{1}{z}\right) R(v,z,x) + \lambda(x)$$

$$\cdot \left[1 + \int_0^1 R(v',z,x) \; w(v') \; dv' \right] \qquad (4.180)$$

$$\cdot \left[1 + \int_0^1 R(v,z',x) \; w(z') \; dz' \right] ,$$

$$e_x(z,x) = -\frac{1}{z} e(z,x)$$

$$+ \left[g(x) + \lambda(x) \int_0^1 e(z',x) \, w(z') \, dz' \right] \quad (4.181)$$

$$\cdot \left[1 + \int_0^1 R(z,u',x) \, w(u') \, du' \right] \; ,$$

$$R(v,z,0) = 0 \; , \quad\quad\quad\quad\quad\quad\quad\quad (4.182)$$

$$e(z,0) = 0 \; . \quad\quad\quad\quad\quad\quad\quad\quad (4.183)$$

At $x = t$ the initial conditions on the functions J and u are

$$J(t,t,z) = \lambda(t) \left[1 + \int_0^1 R(v',z,t) \, w(v') \, dv' \right] \quad (4.184)$$

and

$$u(t,t) = g(t) + \lambda(t) \int_0^1 e(z',t) \, w(z') \, dz' \; . \quad (4.185)$$

For the independent variable x in the interval (t,x_1) the differential equations for the functions J and u are adjoined to those for the functions R and e . They are

$$J_x(t,x,z) = -\frac{1}{z} J(t,x,z) + \lambda(x)$$

$$\cdot \left[1 + \int_0^1 R(v',z,x) \, w(v') \, dv' \right] \quad\quad (4.186)$$

$$\cdot \int_0^1 J(t,x,z') \, w(z') \, dz'$$

and

$$u_x(t,x) = \left[g(x) + \lambda(x) \int_0^1 e(z',x) \, w(z') \, dz' \right]$$

(4.187)

$$\cdot \int_0^1 J(t,x,z') \, w(z') \, dz' \; .$$

4.12   VALIDATION FOR CASE OF GENERAL FORCING FUNCTION

Earlier we showed that the function J determined by the Cauchy system given in the previous section satisfies the Fredholm integral equation

$$J(t,x,z) = \lambda(t) \, e^{-(x-t)/z}$$

(4.188)

$$+ \lambda(t) \int_0^x k(|t-y|) \, J(y,x,z) \, dy \; ,$$

$$0 \le t \le x \le x_1, \quad 0 \le z \le 1 \; .$$

where the kernel k is given in terms of the function w by means of the relation

$$k(r) = \int_0^1 e^{-r/z'} \, w(z') \, dz' \; ,$$

(4.189)

$$r > 0 \; .$$

As a corollary, note that the function Φ, defined by the relation,

$$\Phi(t,x) = \int_0^1 J(t,x,z') \, w(z') \, dz' \; ,$$

(4.190)

$$0 \le t \le x \le x_1 \; ,$$

where the function J is determined by the Cauchy system, satisfies the family of Fredholm integral equations

$$\Phi(t,x) = \lambda(t) \, k(x-t) + \lambda(t) \int_0^x k(|t-y|) \, \Phi(y,x) \, dy \, ,$$

$$0 \le t \le x \le x_1 \, . \quad (4.191)$$

Introduce the function f by means of the equation

$$f(z,x) = \int_0^x e^{-(x-y)/z} \, u(y,x) \, dy, \quad (4.192)$$

$$0 \le x \le x_1, \quad 0 \le z \le 1 \, ,$$

where the function u is determined by the Cauchy system of the previous section. We wish to show that

$$f(z,x) = e(z,x) \, , \quad (4.193)$$

$$0 \le x \le x_1, \quad 0 \le z \le 1 \, .$$

For the function f we see that

$$f_x(z,x) = -\frac{1}{z} \, f(z,x) + u(x,x)$$

$$\quad (4.194)$$

$$+ \int_0^x e^{-(x-y)/z} \, u(x,x) \quad \Phi(y,x) \, dy \, ,$$

or

$$f_x(z,x) = -\frac{1}{z} \, f(z,x) + u(x,x)$$

$$\quad (4.195)$$

$$\cdot \left[1 + \int_0^x e^{-(x-y)/z} \, \Phi(y,x) \, dy \right] .$$

It follows that

$$f_x(z,x) = -\frac{1}{z} f(z,x)$$

$$+ \left[g(x) + \lambda(x) \int_0^1 e(z',x) \ w(z') \ dz' \right] \qquad (4.196)$$

$$\cdot \left[1 + \int_0^x e^{-(x-y)/z} \int_0^1 J(y,x,z') \ w(z') \ dz' \ dy \right],$$

which becomes

$$f_x(z,x) = -\frac{1}{z} f(z,x)$$

$$+ \left[g(x) + \lambda(x) \int_0^1 e(z',x) \ w(z') \ dz' \right] \qquad (4.197)$$

$$\cdot \left[1 + \int_0^1 R(z,z',x) \ w(z') \ dz' \right] .$$

In obtaining the last equation use is made of Eq. (4.31), established earlier. Furthermore the initial condition on the function f is

$$f(z,0) = 0, \qquad\qquad\qquad\qquad\qquad (4.198)$$

$$0 \le z \le 1 .$$

Again comparing Cauchy problems, now for the functions f and e, it is seen that

$$e(z,x) = \int_0^x e^{-(x-y)/z} \ u(y,x) \ dy, \qquad\qquad (4.199)$$

$$0 \le x \le x_1, \qquad 0 \le z \le 1 .$$

Lastly we introduce the function N by means of the definition

$$N(t,x) = g(t) + \lambda(t) \int_0^x k(|t-y|) \, u(y,x) \, dy \, , \qquad (4.200)$$

$$0 \leq t \leq x \leq x_1, \quad 0 \leq t \leq 1 \, ,$$

where the function u is determined by the Cauchy problem in the previous section, and the function k is given by Eq. (4.189) in terms of the function w. We shall show that

$$u(t,x) = N(t,x) \, , \qquad\qquad\qquad (4.201)$$

$$0 \leq t \leq x \leq x_1 \, .$$

Differentiation shows that

$$N_x(t,x) = \lambda(t) \, k(x-t) \, u(x,x)$$

$$+ \lambda(t) \int_0^x k(|t-y|) \, u_y(y,x) \, u(x,x) \, dy \qquad (4.202)$$

$$= u(x,x) \, \Phi(t,x) \, ,$$

where use has been made of Eq. (4.191). This may be rewritten as

$$N_x(t,x) = \left[g(x) + \lambda(x) \int_0^1 e(z',x) \, w(z') \, dz' \right]$$

$$\cdot \int_0^1 J(t,x,z') \, w(z') \, dz' \, . \qquad (4.203)$$

In addition at $x = t$ the function N satisfies the initial condition

$$N(t,t) = g(t) + \lambda(t) \int_0^1 k(t-y) \, u(y,t) \, dy$$

$$= g(t) + \lambda(t) \int_0^t \int_0^1 e^{-(t-y)/z'} w(z') \, dz' \, u(y,t) \, dy$$

$$(4.204)$$

$$= g(t) + \lambda(t) \int_0^1 e(z',t) \, w(z') \, dz' \, ,$$

where Eq. (4.199) plays an essential role. It follows that Eq. (4.201) holds, or

$$u(t,x) = g(t) + \lambda(t) \int_0^x k(|t-y|) u(y,x) \, dy \, , \qquad (4.205)$$

$$0 \leq t \leq x \leq x_1 \, .$$

This completes the proof that the function u , produced by the Cauchy system of the previous section, satisfies the family of Fredholm integral equations in Eq. (4.168).

4.13   THE RESOLVENT KERNEL

An alternative approach to the solution of the integral equation

$$u(t,x) = g(t) + \int_0^x k(|t-y|) \, u(y,x) \, dy \, , \qquad (4.206)$$

$$0 \leq t \leq x \leq x_1 \, ,$$

is through the resolvent kernel K . In terms of this func-tion the solution of Eq. (4.206) may be written as

$$u(t,x) = g(t) + \int_0^x K(t,y,x) \, g(y) \, dy \, , \qquad (4.207)$$

$$0 \le t \le x \le x_1 \, .$$

We shall now obtain a Cauchy system for the resolvent. By differentiation of Eq. (4.207) we see that

$$u_x(t,x) = K(t,x,x) \, g(x) + \int_0^x K_x(t,y,x) \, g(y) \, dy \, . \qquad (4.208)$$

On the other hand we know that

$$u_x(t,x) = \Phi(t,x) \, u(x,x) \qquad (4.209)$$

$$= \Phi(t,x) \left[g(x) + \int_0^x K(x,y,x) \, g(y) \, dy \right] \, .$$

Since the last two equations hold for an arbitrary choice of the function g, it follows that

$$K(t,x,x) = \Phi(t,x) \qquad (4.210)$$

and

$$K_x(t,y,x) = \Phi(t,x) \, K(x,y,x) \, . \qquad (4.211)$$

The function K is known to be symmetric,

$$K(t,y,x) = K(y,t,x) \, , \qquad (4.212)$$

so the differential equation (4.211) may be cast in the form

$$K_x(t,y,x) = \Phi(t,x)\,\Phi(y,x)\ , \tag{4.213}$$

$$x \geq \max(t,y)\ .$$

For definiteness assume that t and y are fixed and such that

$$0 \leq t \leq y \leq x \leq x_1\ . \tag{4.214}$$

We have already seen how to determine the functions X, Y, J, and Φ as solutions of a Cauchy system. At $x = y$ the initial condition on K is

$$K(t,y,y) = \Phi(t,y)\ . \tag{4.215}$$

For $x \geq y$ the Cauchy system for the functions $X(x,z)\ Y(x,z)$, $J(t,x,z)$, $J(y,x,z)$, $\Phi(t,x)$, $\Phi(y,x)$ is augmented by the differential equation (4.213) for the function $K(t,y,x)$ and the initial condition in Eq. (4.215).

The full set of equations for the Cauchy problem is made up of the differential equations,

$$K_x(t,y,x) = \int_0^1 J(t,x,z')\,w(z')\,dz'$$

$$\cdot \int_0^1 J(y,x,z')\,w(z')\,dz'\ , \tag{4.216}$$

$$x \geq t \geq y\ ,$$

$$J_x(t,x,z) = -\frac{1}{z} J(t,x,z)$$

$$(4.217)$$

$$+ X(x,z) \int_0^1 J(t,x,z') \; w(z') \; dz' \; ,$$

$$x \geq t, \quad 0 \leq z \leq 1 \; ,$$

$$X_x(x,z) = Y(x,z) \int_0^1 Y(x,z') \; w(z') \; dz' \; , \qquad (4.218)$$

$$Y_x(x,z) = -\frac{1}{z} Y(x,z) + X(x,z) \int_0^1 Y(x,z') \; w(z') \; dz',$$

$$x \geq 0, \quad 0 \leq z \leq 1 \; , \quad (4.219)$$

and the initial conditions,

$$K(t,y,y) = \int_0^1 J(t,y,z') \; w(z') \; dz' \qquad (4.220)$$

$$J(t,t,z) = X(t,z) \; , \qquad\qquad (4.221)$$

$$0 \leq z \leq 1 \; ,$$

$$X(0,z) = 1 \; , \qquad\qquad (4.222)$$

$$0 \leq z \leq 1 \; ,$$

$$Y(0,z) = 1 \; , \qquad\qquad (4.223)$$

$$0 \leq z \leq 1 \; .$$

4.14   ANOTHER REPRESENTATION FORMULA

Once again consider the integral equation

$$u(t,x) = g(t) + \int_0^x k(|t-y|) \, u(y,x) \, dy \, , \qquad (4.224)$$
$$0 \leq t \leq x \leq x_1 \, .$$

as well as the related integral equation for the function Φ ,

$$\Phi(t,x) = k(x-t) + \int_0^x k(|t-y|) \, \Phi(y,x) \, dy \, , \qquad (4.225)$$
$$0 \leq t \leq x \leq x_1 \, .$$

We shall show that the function u may be represented in terms of the auxiliary function Φ and the forcing function g .

As we know, the function u_x is given by the formula

$$u_x(t,x) = \Phi(t,x) \, u(x,x) \, . \qquad (4.226)$$

Integration then yields the result

$$u(t,x) = u(t,t) + \int_t^x \Phi(t,s) \, u(s,s) \, ds \, . \qquad (4.227)$$

For the function $u(x,x)$ we may write

$$u(x,x) = g(x) + \int_0^x k(x-y) \, u(y,x) \, dy \, , \qquad (4.228)$$

which follows from Eq. (4.224). But from Eqs. (4.224) and (4.225) we see that

$$\int_0^x k(x-y) \; u(y,x) \; dy = \int_0^x \Phi(t,x) \; g(t) \; dt \; , \qquad (4.229)$$

so that Eq. (4.228) becomes

$$u(x,x) = g(x) + \int_0^x \Phi(y,x) \; g(y) \; dy \; , \qquad (4.230)$$
$$0 \le x \le x_1 \; .$$

It follows from Eq. (4.227) and (4.230) that the function u may be represented in the form

$$u(t,x) = g(t) + \int_0^t \Phi(y,t) \; g(y) \; dy$$

$$\qquad\qquad\qquad\qquad\qquad\qquad (4.231)$$

$$+ \int_t^x \Phi(t,s) \left[g(s) + \int_0^s \Phi(y,s) \; g(y) \; dy \right] ds \; ,$$
$$0 \le t \le x \le x_1 \; .$$

By introducing the notation

$$U(x) = g(x) + \int_0^x \Phi(y,x) \; g(y) \; dy \; , \qquad (4.232)$$

Eq. (4.231) may be rewritten as

$$u(t,x) = U(t) + \int_t^x \Phi(t,s) \; U(s) \; ds \; , \qquad (4.233)$$
$$0 \le t \le x \le x_1 \; .$$

Equation (4.233) is a formula for u involving two integrations on the function Φ . Let us derive another
formula, that of Krein.

Let p be the solution of the inhomogeneous integral
equation

$$p(t,x) = 1 + \int_0^x k(|t-y|)\, p(y,x)\, dy ,$$

(4.234)

$$0 \le t \le x \le x_1 .$$

Also observe that

$$\frac{d}{dx} \int_0^x p(t,x)\, g(t)\, dt = p(x,x)\, g(x) + \int_0^x p_x(t,x)\, g(t)\, dt$$

$$= p(x,x) \left[g(x) + \int_0^x \Phi(t,x)\, g(t)\, dt \right]$$

(4.235)

$$= p(x,x)\, u(x,x) ,$$

where we have used the relation

$$p_x(t,x) = \Phi(t,x)\, p(x,x) .$$

(4.236)

Solving Eq. (4.235) for u(x,x) yields

$$u(x,x) = \frac{\frac{d}{dx} \int_0^x p(t,x)\, g(t)\, dt}{p(x,x)}$$

(4.237)

and substitution of this expression into Eq. (4.227) gives us
another representation,

$$u(t,x) = \frac{\dfrac{d}{dt} \displaystyle\int_0^t p(y,t)\, g(y)\, dy}{p(t,t)}$$

$$(4.238)$$

$$+ \int_t^x \frac{p_s(t,s)}{p(s,s)} \; \frac{\dfrac{d}{ds} \displaystyle\int_0^s p(y',s)\, g(y')\, dy'}{p(s,s)} \; ds \; ,$$

$$0 \le t \le x \le x_1 \, .$$

This formula also involves two integrations but it may
be less convenient to use since it involves derivatives
and the whole family of functions $p(y,t)$. There may be
some questions regarding the implementation of Eq. (4.233).
For example, how are the integrals to be evaluated?
The initial value method of this chapter provides an
answer for it is an algorithm for producing $u(t,x)$ via
Eqs. (4.232) and (4.233).

4.15   DISPLACEMENT KERNELS ON THE HALF-LINE

A central functional equation in many branches of applied mathematics -- including optimal filtering and detection, neuronal nets, and astrophysics -- is the integral equation

$$J(t,z) = I \, e^{-t/z} + \int_0^\infty K(|t-y|) \, J(y,z) \, dy , \qquad (4.239)$$

$$0 \leq t < \infty, \quad 0 \leq z \leq 1.$$

The kernel K is an $n \times n$ symmetric square matrix representable in the form

$$K(r) = \int_0^1 e^{-r/z'} \, W(z') \, dz' , \qquad (4.240)$$

$$r > 0,$$

where W is a square $n \times n$ matrix for which

$$W^T(z) = W(z) , \qquad (4.241)$$

$$0 \leq z \leq 1.$$

As usual, the superscript T represents a simple transposition. The matrix I is the unit $n \times n$ matrix, and J is also a square $n \times n$ matrix. We further assume that

$$K(\infty) \, J(\infty,z) = 0 , \qquad (4.242)$$

$$0 \leq z \leq 1.$$

The aim of this paper is to discuss a Cauchy system

which is equivalent to the original Fredholm integral equation.   The Cauchy system is

$$J_t(t,z) = - z^{-1} J(t,z)$$

$$+ \int_0^1 J(t,z') W(z') dz' \ H(z) , \qquad (4.243)$$

$$t \geq 0,$$

$$J(0,z) = H(z) , \qquad\qquad\qquad (4.244)$$

$$0 \leq z \leq 1,$$

and the matrix H satisfies the nonlinear integral equation

$$H(z) = I + \int_0^1 W(z') \frac{z'z}{z'+z} H^T(z') H(z) dz' , \qquad (4.245)$$

$$0 \leq z \leq 1.$$

Though this equivalence is of analytic interest, its
greatest utility may be on the computational side.   The integral equation (4.245) for the auxiliary function H is
readily resolved via successive approximations, and then the
method of lines may be used to handle equations (4.243) and
(4.244).

Let us first show that the solution of the Fredholm
integral equation (4.239) satisfies the Cauchy system in Eqs.
(4.243)-(4.245).   Rewrite the integral equation (4.239) in
the form

$$J(t,z) = I\,e^{-t/z} + \int_0^t K(t-y)\,J(y,z)\,dy$$

$$(4.246)$$

$$+ \int_t^\infty K(y-t)\,J(y,z)\,dy.$$

Then differentiate both sides of this equation with respect to t. The result is

$$J_t(t,z) = -z^{-1}\,I\,e^{-t/z} + \int_0^t K'(t-y)\,J(y,z)\,dy$$

$$(4.247)$$

$$- \int_t^\infty K'(y-t)\,J(y,z)\,dy.$$

The prime in Eq. (4.246), of course, denotes differentiation o of the matrix K with respect to its argument. Through integration by parts the last equation becomes

$$J_t(t,z) = -z^{-1}\,I\,e^{-t/z} - K(t-y)\,J(y,z)\,\Big|_0^t$$

$$+ \int_0^t K(t-y)\,J_t(y,z)\,dy \qquad (4.248)$$

$$- \left[K(y-t)\,J(y,z)\,\Big|_t^\infty - \int_t^\infty K(y-t)\,J_t(y,z)\,dy\right].$$

Upon simplification this last equation is seen to be

$$J_t(t,z) = - z^{-1} I e^{-t/z} + K(t) J(0,z)$$

(4.249)

$$- K(\infty) J(\infty,z) + \int_0^\infty K(|t-y|) J_t(y,z) dy .$$

Finally, using the assumption in Eq. (4.242), it is seen that the matrix J_t satisfies the integral equation

$$J_t(t,z) = - z^{-1} I e^{-t/z} + K(t) J(0,z)$$

(4.250)

$$+ \int_0^\infty K(|t-y|) J_t(y,z) dy ,$$

$$0 \leq t < \infty, \quad 0 \leq z \leq 1.$$

The solution of this integral equation is seen to be

$$J_t(t,z) = - z^{-1} J(t,z)$$

(4.251)

$$+ \left[\int_0^1 J(t,z') W(z') dz' \right] H(z) ,$$

where

$$H(z) = J(0,z) .$$

(4.452)

These are Eqs. (4.243) and (4.244). It remains to derive the integral equation (4.245) for the matrix H.

According to Eq. (4.293) we have

$$J(0,z) = I + \int_0^\infty K(y) J(y,z) dy .$$

(2.253)

From Eq. (4.240) it follows that

$$H(z) = I + \int_0^\infty \int_0^1 e^{-y/z'} W(z') \, dz' \, J(y,z) \, dy$$

(4.254)

$$= I + \int_0^1 W(z') \, R(z',z) \, dz' \, ,$$

where

$$R(v,z) = \int_0^\infty e^{-y/v} \, J(y,z) \, dy \, .$$

(4.255)

Through integration by parts we see that

$$R(v,z) = -v \, J(y,z) \, e^{-y/v} \bigg|_0^\infty$$

(4.256)

$$+ v \int_0^\infty e^{-y/v} \, J_t(y,z) \, dy \, .$$

Employment of the differential equation (4.251) then yields the relation

$$R(v,z) = vJ(0,z) + v \int_0^\infty e^{-y/v}$$

(4.257)

$$\cdot \left\{ z^{-1} \, J(y,z) + \left[\int_0^1 J(y,z') \, W(z') \right] H(z) \right\} \, dy$$

or

$$R(v,z) = vH(z) - vz^{-1} R(v,z)$$
$$+ v\int_0^1 R(v,z') \, W(z') \, dz' \, H(z) \; .$$

(4.258)

To evaluate the integral in the last equation, we note that transposition in Eq. (4.254) yields the relation

$$H^T(z) = I + \int_0^1 R^T(z',z) \, W(z') \, dz \; . \tag{4.259}$$

We shall now show that

$$R^T(z,v) = R(v,z) \; . \tag{4.260}$$

According to Eq. (4.239) we have

$$J(t,z) = I \, e^{-t/z} + \int_0^\infty K(|t-y|) \, J(y,z) \, dy \tag{4.261}$$

and

$$J(t,y) = I \, e^{-t/v} + \int_0^\infty K(|t-y|) \, J(y,v) \, dy \; . \tag{4.262}$$

Eq. (4.262) may be transformed into the relation

$$J^T(t,v) = I \, e^{-t/v} + \int_0^\infty J^T(y,v) \, K(|t-y|) \, dy \; , \tag{4.263}$$

where we have used the assumed symmetry of the matrix K.
Through cross multiplication and integration Eqs. (4.261) and
(4.263) become

$$\int_0^\infty J^T(t,v) \ e^{-t/z} \ dt + \int_0^\infty \int_0^\infty J^T(t,v) \ K(|t-y|) \ J(y,z) dy \ dt$$

$$= \int_0^\infty J(t,z) \ e^{-t/v} \ dt \qquad\qquad (2.264)$$

$$+ \int_0^\infty \int_0^\infty J^T(y,v) \ K(|t-y|) \ J(t,z) \ dy \ dt$$

or

$$\int_0^\infty J^T(t,v) \ e^{-t/z} \ dt = \int_0^\infty J(t,z) \ e^{-t/v} \ dt \ . \qquad (2.265)$$

But this is precisely the desired result contained in Eq.
(4.260). Returning to Eq. (4.258), we find that

$$R(v,z) = vH(z) - v \ z^{-1} \ R(v,z)$$
$$\qquad\qquad\qquad\qquad\qquad\qquad\qquad (2.266)$$
$$+ v \int_0^1 R^T(z',v) \ W(z') \ dz' \ H(z) \ .$$

In view of Eq. (4.259) we have

$$R(v,z) = v \ H(z) - v \ z^{-1} \ R(v,z)$$
$$\qquad\qquad\qquad\qquad\qquad\qquad\qquad (4.267)$$
$$+ v\left[H^T(v) - I\right] H(z) \ .$$

or

$$R(v,z) = \frac{v\,z}{v+z}\,H^T(v)\,H(z) \; . \tag{4.268}$$

Returning to Eq. (4.254), we find the desired integral equation for the function H,

$$H(z) = I + \int_0^1 W(z') \frac{z'\,z}{z'+z}\,H^T(z')\,H(z)\,dz' \; . \tag{4.269}$$

Alternatively, this may be written

$$H^T(z) = I + z\,H^T(z)\int_0^1 \frac{v'}{v'+z}\,H(v')\,W(v')\,dv' \; . \tag{4.270}$$

The proof of necessity is now complete.

We next wish to show that the solution of the Cauchy system in Eqs. (4.243)-(4.245) provides the solution of Eq. (4.239), the kernel being given by Eq. (4.240). Introduce the new function M by means of the relation

$$M(t,z) \equiv I\,e^{-t/z} + \int_0^\infty K(|t-y|)\,J(y,z)\,dy \; , \tag{4.271}$$

$$0 \leq t < \infty, \quad 0 < z \leq 1,$$

where the matrix K is given by Eq. (4.240). Our goal is to show that the function M satisfies the same Cauchy system that the function J does. First, we consider the differential equation for the function M. We write

$$M(t,z) = I\ e^{-t/z} + \int_0^t K(t-y)\ J(y,z)\ dy$$

$$(4.272)$$

$$+ \int_t^\infty K(y-t)\ J(y,z)\ dy$$

and by differentiation obtain

$$M_t(t,z) = -z^{-1}\ I\ e^{-t/z} + \int_0^t K'(t-y)\ J(y,z)\ dy$$

$$(4.273)$$

$$- \int_t^\infty K'(y-t)\ J(y,z)\ dy\ .$$

Through integration by parts the above relation becomes

$$M_t(t,z) = -z^{-1}\ I\ e^{-t/z} - K(t-y)\ J(y,z)\ \Big|_0^t$$

$$+ \int_0^t K(t-y)\ J_t(y,z)\ dy \qquad (4.274)$$

$$- \left[K(y-t)\ J(y,z)\ \Big|_t^\infty - \int_t^\infty K(y-t)\ J_y(y,z)\ dy \right]\ .$$

Through simplification this becomes

$$M_t(t,z) = -z^{-1}\ I\ e^{-t/z} + K(t)\ J(0,z) - K(\infty)\ J(\infty,z)$$

$$+ \int_0^\infty K(|t-y|)\ J_t(y,z)\ dy\ . \qquad (4.275)$$

In view of the differential equation (4.243), we find that

$$M_t(t,z) = -z^{-1} \, I \, e^{-t/z} + K(t) \, H(z)$$

$$+ \int_0^\infty K(|t-y|) \left\{ -z^{-1} \, J(y,z) \right.$$

$$\left. + \int_0^1 J(t,z') \, W(z') \, dz' \, H(z) \right\} \quad dy$$

or

$$M_t(t,z) = -z^{-1} \left[I \, e^{-t/z} + \int_0^\infty K(|t-y|) \, J(y,z) \, dy \right]$$

$$\hspace{9cm} (4.276)$$

$$+ \quad K(t) + \int_0^\infty K(|t-y|) \int_0^1 J(t,z')W(z')dz'dy \, H($$

Keeping in mind the definitions of the functions M and K contained in Eqs. (4.272) and (4.240), it is seen that

$$M_t(t,z) = -z^{-1} \, M(t,z)$$

$$\hspace{9cm} (4.277)$$

$$+ \left[\int_0^1 M(t,z') \, W(z') \, dz' \right] H(z) \, ,$$

$$0 \leq t < \infty, \quad 0 \leq z \leq 1.$$

This establishes that the functions M and J satisfy the same differential equation. Next we must consider the initial condition at $t = 0$.

    Let

$$N(z) = M(0,z) = I + \int_0^\infty K(y) \, J(y,z) \, dy \; . \qquad (4.278)$$

Use of Eq. (4.240) yields

$$N(z) = I + \int_0^\infty \int_0^1 e^{-y/z'} \, W(z') \, dz' \, J(y,z) \, dy$$

$$(4.279)$$

$$= I + \int_0^1 W(z') \, R(z',z) \, dz' \; ,$$

where

$$R(v,z) = \int_0^\infty e^{-y/v} \, J(y,z) \, dy \; , \qquad (4.280)$$

$$0 \le v, z \le 1 .$$

We wish to show that

$$R(v,z) = \frac{vz}{v+z} \, H^T(v) \, H(z) \; . \qquad (4.281)$$

Through integration by parts and employment of Eq. (4.243), we see that

$$R(v,z) = v \, J(y,z) \, e^{-y/v} \, \Big|_0^\infty$$

$$+ v \int_0^\infty e^{-y/v} \, J_t(y,z) \, dy \qquad (4.282)$$

$$= v \, J(0,z) + v \int_0^\infty e^{-y/v} \left\{ -z^{-1} \, J(y,z) + \right.$$

$$+ \int_0^1 J(y,z') \; W(z') \; dz' \; H(z) \Big\} \quad dy \; . \qquad (4.282)$$

This may be rewritten as

$$R(v,z) = v \; H(z) - v \; z^{-1} \; R(v,z)$$
$$\qquad (4.283)$$
$$+ v \int_0^1 R(v,z') \; W(z') \; dz' \; H(z) \; .$$

Introducing the auxiliary matrix A by the definition

$$A(v) - I = \int_0^1 R(v,z') \; W(z') \; dz' \; , \qquad (4.284)$$
$$0 \leq v \leq 1 \; ,$$

it is found that Eq. (4.283) becomes

$$R(v,z) = \frac{vz}{v+z} \; A(v) \; H(z) \; . \qquad (4.285)$$

In view of this, it is seen that the matrix A satisfies the integral equation

$$A(v) = I + \int_0^1 \frac{vz'}{v+z'} \; A(v) \; H(z') \; W(z') \; dz' \; , \qquad (4.286)$$
$$0 \leq v \leq 1 \; .$$

Since the function H satisfies the integral equation, the function H^T satisfies the integral equation

$$H^T(z) = I + \int_0^1 H^T(z) \; H(z') \; \frac{z'z}{z'+z} \; W(z') \; dz' \; . \qquad (4.287)$$

Assuming that Eq. (4.286) has a unique solution we see that

$$A(v) = H^T(v) \; , \qquad\qquad\qquad (4.288)$$
$$0 \le v \le 1 .$$

Eq. (4.285) then becomes

$$R(v,z) = \frac{vz}{v+z} \; H^T(v) \; H(z) \; . \qquad\qquad (4.289)$$

It follows, finally, that

$$N(z) = I + \int_0^1 W(z') \; R(z',z) \; dz'$$
$$\qquad\qquad\qquad\qquad\qquad (4.290)$$
$$= I + \int_0^1 W(z') \; \frac{z'z}{z'+z} \; H^T(z') \; H(z) \; dz' \; .$$

In view of the integral equation (4.245) for the function H,

$$N(z) = H(z) \; , \qquad\qquad\qquad\qquad (4.291)$$
$$0 \le z \le 1 ,$$

which is the desired initial condition on the function M. This completes the proof that the solution of the Cauchy system in Eqs. (4.243)-(4.245) provides the solution of the Fredholm integral equation (4.239).

Another important function, related to the resolvent matrix, is the function Φ which is defined to be the solution of the matrix integral equation

$$\Phi(t) = K(t) + \int_0^\infty K(|t-y|)\ \Phi(y)\ dy\ , \qquad (4.292)$$

$$0 \leq t < \infty.$$

According to Eqs. (4.239) and (4.240), we see that the function Φ may be represented in the form

$$\Phi(t) = \int_0^1 J(t,z')\ W(z')\ dz'\ , \qquad (4.293)$$

$$0 \leq t < \infty.$$

Observe then that Eq. (4.243) may be rewritten in the form

$$J_t(t,z) = -z^{-1}\ J(t,z) + \Phi(t)\ H(z)\ . \qquad (4.294)$$

Keeping in mind the initial condition in Eq. (4.244), we find that

$$J(t,z) = \left[I\ e^{-t/z} + \int_0^t e^{-(t-y)/z}\ \Phi(y)\ dy \right] H(z)\ . \qquad (4.295)$$

By multiplying both sides of this equation on the right by W and integrating we find that the function Φ satisfies the Volterra integral equation

$$\Phi(t) = L(t) + \int_0^t \Phi(y) \, L(t-y) \, dy \, ,\qquad (4.296)$$

$$t \geq 0,$$

where

$$L(t) = \int_0^1 e^{-t/z'} \, H(z') \, W(z') \, dz' \, ,\qquad (4.297)$$

$$t \geq 0.$$

It follows that the solution J of the Cauchy system provides a solution of the Volterra equation in Eqs. (4.296) and (4.297) through the representation formula Eq. (4.293).

4.16   AN INTEGRAL EQUATION OF RAREFIED GAS DYNAMICS

The integral equation for the function G,

$$\pi^{1/2} \, G(\eta) = J_1(\eta) - J_1(\alpha-\eta)\qquad (4.298)$$

$$+ \int^{\alpha/2} \left[J_{-1}(|\eta-\eta'|) - J_{-1}(\alpha-\eta-\eta') \right] G(\eta') \, dz' \, ,$$

$$0 \leq \eta \leq \alpha/2,$$

plays a significant role in the theory of rarefied gas dynamics. The functions J_1 and J_{-1}, not to be confused with our earlier notation, are defined by the relation

$$J_n(\eta) = \int_0^\infty y^n e^{-y^2} \, e^{-\eta/y} \, dy \, ,\qquad (4.299)$$

$$\eta > 0, \quad n = +1.$$

On the other hand, earlier in this chapter a new theory for
the solution of Fredholm integral equations of the form

$$u(t) = g(t) + \int_0^C k(|t-y|) \, u(y) \, dy , \qquad (4.300)$$
$$0 < t < C ,$$

has been presented. This theory converts Eq. (4.300) into
an initial value problem for a system of ordinary differential equations. Its solution is admirably suited for modern
analog and digital computers. This section applies the
initial value method to Eq. (4.298), which leads to some
new equations. It also describes the results of numerical
experiments.

The integral equation for G may be written

$$\pi^{1/2} G(\eta) = J_1(\eta) - J_1(\alpha-\eta) + \int_0^{\alpha/2} J_{-1}(|\eta-\eta'|) G(\eta')d\eta'$$

$$- \int_0^{\alpha/2} J_{-1}(\alpha-\eta-\eta') G(\eta')d\eta',$$
$$0 \leq \eta \leq \alpha/2. \quad (4.301)$$

Furthermore, the function G has the anti-symmetric property

$$G(\eta) = -G(\alpha-\eta) , \qquad (4.302)$$
$$0 \leq \eta \leq \alpha/2.$$

By introducting the new variable of integration η'' ,

$$\eta'' = \alpha - \eta' , \qquad (4.303)$$

the last integral in Eq. (4.301) may be written

$$-\int_0^{\alpha/2} J_{-1}(\alpha-\eta-\eta') \, G(\eta') \, d\eta'$$

(4.304)

$$= \int_{\alpha/2}^{\alpha} J_{-1}(|\eta''-\eta|) \, G(\eta'') \, d\eta'' \, .$$

Thus the integral equation for the function G assumes the form

$$\pi^{1/2} \, G(\eta) = J_1(\eta) - J_1(\alpha-\eta) + \int_0^{\alpha} J_{-1}(|\eta-\eta'|) G(\eta') \, d\eta' \, ,$$

$$0 \le \eta \le \alpha.$$

(4.305)

Let us now rederive an initial value problem for the integral equation

$$I(t,x,z) = e^{-(x-t)/z} + \int_0^x k(|t-y|) \, I(y,x,z) \, dy \, ,$$

$$0 \le t \le x.$$

(4.306)

After this is accomplished, by making use of the definition of $J_n(\eta)$ in Eq. (4.299), the solution to the desired integral equation may be obtained by forming suitable linear combinations of solutions to Eq. (4.306).

Consider the integral equation

$$I(t,x,z) = e^{-(x-t)/z} + \int_0^x k(|t-y|) \, I(y,x,z) \, dy \, ,$$

$$0 \le t \le x, \quad a \le z \le b,$$

$$(4.307)$$

where the kernel is expressed in the form

$$k(r) = \int_a^b e^{-r/z'} \, w(z') \, dz' \, , \qquad (4.308)$$

a slight generalization of our earlier assumption. It is desired to find an initial value problem satisfied by the function .

By differentiating both sides of Eq. (4.307) with respect to x, one can see that

$$I_x(t,x,z) = -z^{-1} e^{-(x-t)/z} + k(x-t) \, I(x,x,z)$$

$$+ \int_0^x k(|t-y|) \, I_x(y,x,z) \, dy \, . \qquad (4.309)$$

This is an integral equation for the function I_x. Its solution is

$$I_x(t,x,z) = -z^{-1} I(t,x,z) + I(x,x,z) \, \Phi(t,x) \, , \qquad (4.310)$$

where the function $\Phi(t,x)$ has been introduced as the solution to the integral equation

$$\Phi(t,x) = k(x-t) + \int_0^x k(|t-y|) \; \Phi(y,x) \; dy \;, \qquad (4.311)$$

$$0 \le t \le x.$$

Note that according to Eqs. (4.307) and (4.308), the function $\Phi(t,x)$ may be expressed as

$$\Phi(t,x) = \int_a^b I(t,x,z') \; w(z') \; dz' \; . \qquad (4.312)$$

In view of this relation, all that remains is to consider the function $I(x,x,z)$. First, though, note that

$$I_x(0,x,z) = -z^{-1} \; I(0,x,z) + I(x,x,z) \; \Phi(0,x) \; . \qquad (4.313)$$

In the integral equation (4.307) replace t by $x-t$, which results in the relation

$$I(x-t,x,z) = e^{-t/z} + \int_0^x k(|x-t-y|) \; I(y,x,z) \; dy \; . \qquad (4.314)$$

Now introduce the new variable of integration y' as

$$y' = x - y \; . \qquad (4.315)$$

Equation (4.314) takes the form

$$I(x-t,x,z) = e^{-t/z} + \int_0^x k(|y'-t|) \; I(x-y',x,z) \; dy \; . \qquad (4.316)$$

Differentiate Eq. (4.316) with respect to x to obtain

$$\frac{d}{dx} I(x-t,x,z) = k(x-t) \ I(0,x,z)$$

$$+ \int_0^x k(|y'-t|) \ \frac{d}{dx} I(x-y',x,z) \ dy' \ .$$

$$(4.317)$$

The solution of this integral equation is

$$\frac{d}{dx} I(x-t,x,z) = I(0,x,z) \ \Phi(t,x) \ . \tag{4.318}$$

For t = 0, this equation becomes

$$\frac{d}{dx} I(x,x,z) = I(0,x,z) \ \Phi(0,x) \ . \tag{4.319}$$

Equations (4.313) and (4.318), together with Eq. (4.312), constitute a set of differential equations for the functions $I(0,x,z)$ and $I(x,x,z)$. Let us write

$$X(x,z) = I(x,x,z) \ , \tag{4.320}$$

$$Y(x,z) = I(0,x,z) \ , \tag{4.321}$$

$$a \leq z \leq b.$$

The functions X and Y satisfy the differential equations

$$X_x(x,z) = Y(x,z) \int_b^a Y(x,a') \ w(z') \ dz' \tag{4.322}$$

and

$$Y_x(x,z) = -z^{-1} Y(x,z) + X(x,z)\int_a^b Y(x,z')w(z') \, dz' .$$

$$(4.323)$$

The initial conditions at $x = 0$ can be read off from Eq. (4.307); they are

$$X(0,z) = 1 , \qquad\qquad\qquad (4.324)$$

$$Y(0,z) = 1 , \qquad\qquad\qquad (4.325)$$

$$a \leq z \leq b.$$

These equations determine the important initial and terminal values of I without solving the integral equation, Eq. (4.307).

For a fixed positive value of t, Eqs. (4.322) and (4.323) -- subject to the initial conditions in Eqs. (4.324) and (4.325) -- determine X and Y (assuming the solution exists and is unique). The values of $X(t,z)$, $a \leq z \leq b$, provide the initial condition for $I(t,x,z)$ at $x = t$,

$$I(t,t,z) = X(t,z) . \qquad\qquad\qquad (4.326)$$

Then the differential equations (Eqs. (4.322) and (4.323)) for X and Y, together with the equation for I,

$$I_x(t,x,z) = -z^{-1} I(t,x,z)$$

$$+ X(x,z)\int_a^b I(t,x,z') \, w(z') \, dz' , \qquad (4.327)$$

determine I for x≥t.   In addition, Eq. (4.312) determines
the function Φ for t≤x, completing the derivation of the
initial value problem for the function I.

The problem at hand has the form

$$\pi^{1/2} \, G(\eta) = J_1(\eta) - J_1(\alpha - \eta)$$

$$\hspace{6cm} (4.328)$$

$$+ \int_0^\alpha J_{-1}(|\eta - \eta'|) \, G(\eta') \, d\eta' \; .$$

To fit the solution of Eq. (4.328) into the context of the
initial value problem just derived, a few minor transformations are necessary.

Note first that the function w(z), which defines
the kernel in Eq. (4.308), takes the form

$$w(z) = \pi^{-1/2} z^{-1} e^{-z^2} \; , \hspace{3cm} (4.329)$$

$$a = 0 \leq z < \infty = b \; ,$$

when Eq. (4.328) is under consideration. To eliminate the
inconvenience of an infinite interval, the substitution
$p = e^{-z}$, $0 \leq p \leq 1$, will be made.

Let

$$\xi(x,p) = X(x, -\log p) \; , \hspace{3cm} (4.330)$$

$$\Psi(x,p) = Y(x, -\log p) \; , \hspace{3cm} (4.331)$$

$$\gamma(t,x,p) = (t, x, -\log p) \; , \hspace{3cm} (4.332)$$

and

$$\zeta(t,x,p) = I(x-t,x,-\log p) ,\qquad\qquad (4.333)$$
$$0\le p\le 1.$$

Using Eq. (4.329), the previous initial value problem in Eqs. (4.322)-(4.327) becomes

$$\xi_x(x,p) = -\pi^{-1/2}\Psi(x,p)\int_0^1 \Psi(x,r)\ r^{-(1+\log r)}dr/\log r,$$

$$(4.334)$$

$$\Psi_x(x,p) = \Psi(x,p)/\log p$$

$$(4.335)$$

$$-\pi^{-1/2}\xi(x,p)\int_0^1 \Psi(x,r)\ r^{-(1+\log r)}dr/\log r,$$

$$x\ge 0;$$

$$\gamma_x(t,x,p) = \gamma(t,x,p)/\log p$$

$$(4.336)$$

$$-\pi^{-1/2}\xi(x,p)\int_0^1 \gamma(t,x,r)r^{-(1+\log r)}dr/\log r,$$

and

$$\zeta_x(t,x,p) = -\pi^{-1/2}\Psi(w,p)\int_0^1 \gamma(t,x,r)r^{-(1+\log r)}dr/\log r,$$

$$(r.337)$$

$$x\ge t.$$

The initial conditions are:

$$\xi(0,p) = 1 ,\qquad\qquad (4.338)$$

$$\Psi(0,p) = 1 ,\qquad\qquad (4.339)$$

$$\gamma(t,t,p) = \xi(t,p) \, , \tag{4.340}$$

and

$$\zeta(t,t,p) = \bar{\Psi}(t,p) \, , \tag{4.341}$$

$$0 \leq p \leq 1.$$

Using γ and ζ, the solution of Eq. (4.328) may be expressed as

$$G(\eta,\alpha) = \pi^{-1/2} \int_0^1 \{\zeta(\eta,\alpha,p) - \gamma(\eta,\alpha,p)\} \, w_1(p) \, dp \, , \tag{4.342}$$

where

$$w_1(p) = -p^{-(1+\log p)} \log p \, , \tag{4.343}$$

$$1 \geq p \geq 0.$$

To deal with the equations of the previous paragraphs numerically, we must reduce all transcendental operations to those that can be carried out arithmetically. All integrals will be replaced by finite sums of the form

$$\int_0^1 f(x) \, dx \cong \sum_{i=1}^{N} f(p_i) \, w_i \, , \tag{4.344}$$

where $p_1, p_2, \ldots, N$ are suitable abscissas and $w_1, w_2, \ldots, w_N$ are appropriate weights. In the calculations to be presented, Gaussian quadrature of the order $N = 9$ is employed.

Let us use the notation

$$\xi_i(x) = \xi(x,p_i) , \tag{4.345}$$

$$\Psi_i(x) = \Psi(x,p_i) , \tag{4.346}$$

$$\gamma_i(t,x) = \gamma(t,x,p_i) , \tag{4.347}$$

and

$$\zeta_i(t,x) = \zeta(t,x,p_i) , \tag{4.348}$$

$$i=1,2,\ldots,N.$$

With this convention the initial value problem becomes

$$\xi_i'(x) = -\pi^{-1/2} \Psi_i(x) \sum_{j=1}^{N} \Psi_j(x) p_j^{-(1+\log p_j)} w_j/\log p_j ,$$

$$\tag{4.349}$$

$$\Psi_i'(x) = \Psi_i(x)/\log p_i$$

$$-\pi^{-1/2} \xi_i(x) \sum_{j=1}^{N} \Psi_j(x) p_j^{-(1+\log p_j)} w_j/\log p_j ,$$

$$\tag{4.350}$$

$$x \geq 0;$$

$$\gamma_i'(t,x) = \gamma_i(t,x)/\log p_i$$

$$\tag{4.351}$$

$$-\pi^{-1/2} \xi_i(x) \sum_{j=1}^{N} \gamma_j(t,x) p_j^{-(1+\log p_j)} w_j/\log p_j ,$$

and

$$\zeta_i(t,x) = -\pi^{-1/2}\psi_i(t,x) \sum_{j=1}^{N} \gamma_j(t,x)p_j^{-(1+\log\ p_j)}w_j/\log$$

(4.352)

$$i=1,2,\ldots,N, \quad x \geq t.$$

The initial conditions are

$$\xi_i(0) = 1 \ ,$$

(4.353)

$$\psi_i(0) = 1 \ ,$$

(4.354)

$$\gamma_i(t,t) = \xi_i(t) \ ,$$

(4.355)

and

$$\zeta_i(t,t) = \psi_i(t) \ ,$$

(4.356)

$$i=1,2,\ldots,N.$$

The computational scheme just outlined was applied to Eq. (4.298) with excellent results by J. Casti. The first experiment involved an integration step size of .005 with an Adams-Moulton 4th order integration procedure and a quadratur scheme of $N = 9$ points. Table 4.1 summarizes the results. (We use the notation, .48428 E -6 = .48428 $\times$ 10^{-6}.) The results of Table 4.1 agree to 3 places with those reported in an earlier paper by D. R. Willis.

The accuracy of the results may be indicated by noticing that the function $G(\eta)$ is an odd function about the point $\eta = \alpha/2$. Thus, for a given α, at $\eta = \alpha/2$ the solution should be 0. Our results give values at these points of approximately 10^{-5}. The aforementioned anti-symmetric property of $G(\eta)$ also checks the accuracy of the calculations. Agreement occurs to 5 places. The computing time necessary to generate the results of Table 4.1 on a CDC 6600 computer is about two minutes. Note that the solution is available for all values of α between 0 and 10 that are integral multiples of the integration step size.

To provide another check on a portion of the calculations, 7,9, and 15-point Gaussian quadrature formulas calculate the forcing function $J_1(x)$, $0 \leq x \leq 1$. Table 4.2 summarizes these calculations and compares the results against standard values.

The last experiment examines the changes that take place when a five-point formula and a larger integration step size (.01) are employed. This saves core storage and time by utilizing fewer integration steps. Table 4.3 presents the results at the mid-point $\alpha/2$, while Table 4.4 gives the results for a wider range of values of η and α.

Table 4.3 indicates that the larger step size and coarser quadrature formula maintain 2 to 3-digit accuracy. Computer running time for this case is about .3 minutes on a CDC 6600.

The preceding argument presents an initial value problem whose solution solves a Fredholm integral equation arising in rarefied gas dynamics. Note that during the development of the initial value procedure it was necessary to intorduce the X and Y functions. These functions represented the solution to the integral equation at the end-points of the interval. During an investigation, should one desire only the solution of the integral Eq. (4.239) at the end-points, then a significant saving in computational burden and

Table 4.1

G(η) vs. η FOR VARIOUS VALUES OF α

η \ α	1	2	3	4	5
0	.24813	.35869	.42567	.47107	.50394
.025	.24354	.36293	.43530	.48436	.51989
.050	.23453	.36075	.43729	.48920	.52680
.075	.22401	.35636	.43666	.49112	.53059
.100	.21271	.35083	.43465	.49154	.53276
.150	.18867	.33757	.42804	.48947	.53399
.200	.16332	.32228	.41899	.48469	.53233
.275	.12373	.29682	.40236	.47413	.52620
.350	.08300	.26933	.38325	.46080	.51709
.425	.04164	.24047	.36243	.44555	.50592
.500	.48428E-6	.21063	.34035	.42889	.49323
.625	-.06926	.15938	.30154	.39880	.46959
.750	-.13708	.10686	.26090	.36661	.44364
.875	-.20089	.05360	.21897	.33288	.41602
1.000	-.24813	.67055E-6	.17612	.29801	.38712
1.250		-.10686	.08863	.22583	.32657
1.500		-.21063	.81583E-6	.15154	.26350
1.750		-.30558	.08863	.07604	.19880
2.000		-.35869	.17612	.90525E-6	.13305
2.250			.26090	-.07604	.06667
2.500			.34035	-.15154	-.96485E-6
3.000			.42567	-.29801	-.13305
3.500				-.42888	-.26350
4.000				-.47107	-.38712
5.000					-.50394

Table 4.1 Continued

α / η	6	7	8	9	10
0	.52884	.54835	.56404	.57693	.58769
.025	.54681	.56790	.58486	.59879	.61043
.050	.55529	.57762	.59557	.61031	.62264
.075	.56049	.58393	.60278	.61826	.63119
.100	.56399	.58848	.60817	.62434	.63786
.150	.56775	.59421	.61549	.63297	.64757
.200	.56846	.59678	.61956	.63827	.65391
.275	.56570	.59666	.62158	.64204	.65915
.350	.55981	.59330	.62026	.64240	.66091
.425	.55175	.58770	.61663	.64039	.66026
.500	.54209	.58043	.61128	.63663	.65782
.625	.52337	.56558	.59956	.62749	.65083
.750	.50221	.54820	.58524	.61568	.64113
.875	.47928	.52897	.56900	.60191	.62943
1.000	.45499	.50833	.55131	.58665	.61621
1.250	.40346	.46395	.51273	.55286	.58642
1.500	.34918	.41668	.47115	.51598	.55349
1.750	.29306	.36745	.42753	.47699	.51840
2.000	.23567	.31683	.38244	.43651	.48178
2.250	.17741	.26522	.33631	.39492	.44402
2.500	.11857	.21290	.28939	.35251	.40502
3.000	-.10207E-5	.10689	.19396	.26598	.32642
3.500	-.11857	-.1050E-5	.09796	.17799	.24585
4.000	-.23567	-.10689	.10654E-5	.08918	.16434
5.000	-.45499	-.31682	-.19396	-.08917	.11027E-5

Table 4.2

CALCULATION OF $J_1(x)$ BY QUADRATURE

x	N=7	N=9	N=15	Standard Values
0	.499687	.499963	.500000	.5000
.1	.426092	.426317	.426339	.4263
.2	.369498	.369718	.369766	.3697
.3	.323526	.323777	.323816	.3238
.4	.285281	.285519	.285552	.2855
.5	.252936	.253143	.253176	.2531
.6	.225254	.225433	.225468	.2255
.7	.201349	.201509	.201546	.2015
.8	.180560	.180709	.180746	.1807
.9	.162376	.162517	.162554	.1626
1.0	.146393	.146527	.146563	.1466

Table 4.3

$G(\eta)$ VS. η AT THE POINT $\eta = \alpha/2$ FOR $N = 5$ AND $\Delta = 0.01$

α	1	2	3	4
$\eta = \alpha/2$	.849366E-6	.107288E-5	.125542E-5	.138208E-5

α	5	6	7	8	9	10
$\eta = \alpha/2$	.147521E-5	.154972E-5	.160933E-5	.166148E-5	.169873E-5	.172108E-5

Table 4.4

G(η) VS. η FOR COARSER GRID (N = 5, Δx = .01)

η \ α	1	2	3	4	5
.5	.849366 E-06	.21164	.34284	.43288	.49871
1.0	-.24881	.10729 E-05	.17783	.30152	.39251
1.5		-.21164	.12554 E-05	.15340	.26739
2.0		-.36036	-.17783	.13821 E-05	.13507
2.5			-.34284	-.15340	.14752 E-05
3.0			-.42827	-.30152	-.13507
3.5				-.43287	-.26739
4.0				-.47458	-.39250
4.5					-.49871
5.0					

Table 4.4 Continued

η \ α	6	7	8	9	10
.5	.54893	.58845	.62033	.64660	.66863
1.0	.46217	.51711	.56150	.59810	.62880
1.5	.35515	.42458	.48077	.52715	.56606
2.0	.23991	.32324	.39086	.44673	.49364
2.5	.12078	.21744	.29614	.36130	.41608
3.0	.15497 E-05	.10925	.19869	.27294	.33543
3.5	-.12078	.16093 E-05	.09969	.18279	.25288
4.0	-.23991	-.10924	.16615 E-05	.09163	.16915
4.5	-.35514	-.21743	-.09969	.16987 E-05	.08473
5.0	-.46217	-.32324	-.19868	-.09162	.17211 E-05

cost can be made using only the equations for X and Y.
Thus, to obtain the solution at the boundaries, knowing what
is happening in the interior is not necessary. This is an
important aspect of the imbedding formulation. Putting
$\eta = 0$ and $\eta = \alpha$ in Eq. (4.342), the exact formulas are
seen to be

$$G(0,\alpha) = \pi^{-1/2} \int_0^1 [\xi(\alpha,p) - \Psi(\alpha,p)] \, w_1(p) \, dp$$

$$(4.357)$$

and

$$G(\alpha,\alpha) = -\pi^{-1/2} \int_0^1 [\xi(\alpha,p) - \Psi(\alpha,p)] \, w_1(p) \, dp$$

$$(4.358)$$

where the function w_1 is given by Eq. (4.343). The functions ξ and Ψ are calculated in the Cauchy problem in
Eqs. (4.334), (4.335), (4.338), and (4.339).

4.17   DISCUSSION

This has been a rather lengthy chapter. The theory
for integral equations with displacement kernels is marvel-
ously rich. In addition to the equivalence between this
type of integral equation and a Cauchy system involving an
auxiliary function $R(v,z,x)$, theory provides a reduction
involving auxiliary functions $X(x,z)$ and $Y(x,z)$ when
$\lambda(t) \equiv 1.0$. Furthermore, when in the latter situation,
there are functions called b and h in terms of which all

of the other interesting functionals on the solution can be
algebraically expressed. These b and h functions themselves satisfy a Cauchy system as well as a system of singular integral equations.

For the case of a general forcing function, a Cauchy
system can be both derived and validated. In addition, one
may compute the resolvent and express the solution for arbitrary forcing functions as a quadrature on the resolvent.
Still other alternatives are provided by representation
formulas involving the function $\Phi(t,x)$.

An application of this theory to radiative transfer
is the topic of Chapter 9. Numerous results of computations
are presented for functions of physical interest. Inverse
problems, in which the task is to estimate parameters based
on observations of solutions, are treated in Chapter 9.

The theory also finds application in statistical communication theory, although it is not covered in this book.

In the event that the kernel is of more complex form,
one may wish to study the theory for composite kernels in
the next chapter, and the section in Chapter 9 which deals
with a reduction of dimensionality involving X and Y
functions for a slightly different kernel. It is exciting
to see the presence of X and Y functions in other
settings.

Imbedding concepts of integral equations of displacement type were developed as early as the 1940s by
N. Levison who, while taking an algebraic approach, im-
bedded the solution in a class of solutions of varying interval length.

EXERCISES

1. Define the "transmission function" T,

    $$T(v,u,x) = \int_0^x J(y,x,u) \, e^{-y/v} \, dy \,, \quad \text{and derive}$$

    an initial value problem for T.

2. Derive the initial conditions for b and h.

3. Adapt the results of this chapter to the case,

    $$k(t,y) = e^{-|t-y|} \,, \quad \text{and write a computer program}$$

    for it.

4. A program is given in J. Casti and R. Kalaba, *Imbedding Methods in Applied Mathematics*, Addison-Wesley, 1973, for integral equations with displacement kernels. Punch the program and reproduce some of the results given in Chapter 9 of this book.

REFERENCES

1. H. Kagiwada and R. Kalaba, "An Initial Value Method for Fredholm Integral Equations of Convolution Type," *International Journal of Computational Math.*, Vol. 2 (1968), pp. 143-155.

2. H. Kagiwada, R. Kalaba and A. Schumitzky, "A Representation for the Solution of Fredholm Integral Equations," *Proc. Am. Math. Soc.*, Vol. 23, No. 1 (1969), pp. 37-40.

3. D. R. Willis, "Comparison of Kinetic Theory Analyses of Linearized Couette Flow," *Physics of Fluids*, Vol. 5, No. (1962).

4. H. Kagiwada and R. Kalaba, "The Basic Functions b and
    h for Fredholm Integral Equations with Displacement
    Kernels," *Journal of Optimiz. Theory Appl.*, Vol. 11,
    No. 5 (1973), pp. 517-532.

5. H. Kagiwada and R. Kalaba, "An Initial Value Method Suit-
    able for the Computation of Certain Fredholm Resolvents,"
    Journal of Mathematical and Physical Sciences, Vol. 1,
    No. 2 (1967), pp. 109-122.

6. N. Levinson, "The Wiener RMS Error Criterion in Filter
    Design and Prediction," *Journal of Mathematics and Physics*,
    Vol. 25, No. 4 (1947), pp. 261-278.

7. M. G. Krein, "On a New Method for Solving Linear Integral
    Equations of the First and Second Kinds," *Dokl. Nauk SSSR*,
    Vol. 100 (1955), pp. 413-416.

8. J. Casti, R. Kalaba and V. K. Murthy, "A New Initial Value
    Method for On-line Filtering and Estimation," *IEEE Trans.
    Information Theory*, July 1972, pp. 515-517.

9. T. W. Mullikin, "A Non-linear Integrodifferential Equation
    in Radiative Transfer," *Society for Industrial and Ap-
    plied Mathematics*, Vol. 13 (1965), pp. 388-410.

10. W. Nelson, "Existence, Uniqueness, and Stability of
    Solutions to Chandrasekhar's Integrodifferential Equation
    for X and Y Functions," *Journal of Mathematical
    Analysis and Applications*, Vol. 37 (1972), pp. 580-606.

CHAPTER 5

COMPOSITE KERNELS

5.1   INTRODUCTION

The study of many processes in mathematical physics requires the solution of integral equations of the form

$$u(t) = g(t) + \int_0^c [k_1(|t-y|) + k_2(t+y)] \, u(y) \, dy \, ,$$

$$0 \leq t \leq c.$$

This equation may be treated using the theory of functions of complex variables.   In this chapter, it is shown that this problem may be reduced to the solution of an initial-value problem in a manner which follows naturally from the work on displacement kernels in the preceding chapter.

5.2   DERIVATION OF CAUCHY PROBLEM

Consider the integral equation

$$u(t) = g(t) + \int_0^x k(t,y) \, u(y) \, dy \, , \tag{5.1}$$

$$0 \le t \le x.$$

Assume that the kernel k may be written in the form

$$k(t,y) = k_1(|t-y|) + k_2(t+y) \, , \tag{5.2}$$

$$0 \le t, y,$$

and that

$$k_1(s) = \int_0^1 e^{-s/z'} \, w(z') \, dz' \, , \tag{5.3}$$

$$s \ge 0,$$

$$k_2(s) = \int_0^1 e^{-s/z'} \, w(z') \, r(z') \, dz' \, , \tag{5.4}$$

$$s \ge 0.$$

To indicate the dependence of the solution function u on the length of the interval, write Eq. (5.1) in the form

$$u(t,x) = g(t) + \int_0^x k(t,y) \, u(y,x) \, dy \, , \tag{5.5}$$

$$0 \le t \le x, \quad x \le x_1,$$

and

$$u = u(t,x) \, . \tag{5.6}$$

Differentiate both sides of Eq. (5.5) with respect to x to obtain

$$u_x(t,x) = k(t,x) \, u(x,x) + \int_0^x k(t,y) \, u_x(y,x) \, dy \ .$$

$$(5.7)$$

The solution of this integral equation for u_x is

$$u_x(t,x) = \Phi(t,x) \, u(x,x) \ , \tag{5.8}$$

where the function Φ satisfies the integral equation

$$\Phi(t,x) = k(t,x) + \int_0^x k(t,y) \, \Phi(y,x) \, dy \ , \tag{5.9}$$

$$0 \le t \le x.$$

In view of Eqs. (5.2) and (5.3), and (5.4), introduce the function J,

$$J = J(t,x,z) \ , \tag{5.10}$$

as the solution of the integral equation

$$J(t,x,z) = e^{-(x-t)/z} + r(z) \, e^{-(x+t)/z}$$

$$+ \int_0^x k(t,y) \, J(y,x,z) \, dy \ , \tag{5.11}$$

$$0 \le t \le x.$$

From the superposition principle it follows that

$$\Phi(t,x) = \int_0^1 J(t,x,z') \, w(z') \, dz' \ . \tag{5.12}$$

We shall now consider the determination of the function J as the solution of an initial value method. Then we shall

turn to $u(x,x)$, the second factor on the right-hand side of
Eq. (5.8).

Through differentiation of Eq. (5.11) with respect to
x, we see that

$$J_x(t,x,z) = -\frac{1}{z}\left[e^{-(x-t)/z} + r(z)\ e^{-(x+t)/z}\right]$$

$$+ k(t,x)\ J(x,x,z)$$

$$+ \int_0^x k(t,y)\ J_x(y,x,z)\ dy\ . \qquad (5.13)$$

Viewed as an integral equation for J_x, the solution of Eq.
(5.13) is

$$J_x(t,x,z) = -\frac{1}{z}\ J(t,x,z) + \Phi(t,x)\ J(x,x,z)\ . \qquad (5.14)$$

From Eq. (5.11) it is seen that

$$J(x,x,z) = 1 + r(z)e^{-2x/z} + \int_0^x k(x,y)\ J(y,x,z)\ dy\ . \tag{5.15}$$

The integral in this equation may be written

$$\int_0^x k(x,y)\ J(y,x,z)\ dy$$

$$= \int_0^x \int_0^1 w(z')\left[e^{-(x-y)/z'} + r(z')\ e^{-(x+y)/z'}\right]$$

$$\cdot\ J(y,x,z)\ dy\ dz' \tag{5.16}$$

$$= \int_0^1 w(z') \ R(z',z,x) \ dz' \ , \tag{5.16}$$

where

$$R(v,z,x) = \int_0^x \left[e^{-(x-y)/v} + r(v)e^{-(x+y)/v} \right] J(y,x,z) \ dy \ , \tag{5.17}$$

$$0 \leq v,z \leq 1, \quad 0 \leq x.$$

A Cauchy problem for the function R will now be derived. Differentiation of Eq. (5.17) yields

$$R_x(v,z,x) = -\frac{1}{v} R(v,z,x) + \left[1 + r(v) \ e^{-2x/v} \right] J(x,x,z)$$

$$+ \int_0^1 \left[e^{-(x-y)/v} + r(v) \ e^{-(x+y)/v} \right] J_x(y,x,z) \ dy. \tag{5.18}$$

Use of Eq. (5.14) shows that

$$R_x(v,z,x) = -\left(\frac{1}{v} + \frac{1}{v} \right) R(v,z,x)$$

$$+ \left[1 + r(v) \ e^{-2x/v} \right] J(x,x,z)$$

$$+ J(x,x,z) \int_0^x \left[e^{-(x-y)/v} \right. \tag{5.19}$$

$$+ \left. r(v) \ e^{-(x+y)/v} \right] \Phi(y,x) \ dy \ .$$

The integral in this last equation is evaluated using Eq. (5.12):

$$\int_0^x \left[e^{-(x-y)/v} + r(v) \; e^{-(x+y)/v} \right] \int_0^1 J(y,x,z') \; w(z') \; dz'$$

$$= \int_0^1 R(v,z',x) \; w(z') \; dz' \; . \qquad (5.20$$

The differential equation for R becomes

$$R_x(v,z,x) = - \left(\frac{1}{v} + \frac{1}{z} \right) R(v,z,x) + J(x,x,z)$$

$$\cdot \left[1 + r(v) \; e^{-2x/v} \right. \qquad (5.21$$

$$\left. + \int_0^1 R(v,z',x) \; w(z') \; dz' \right] \; .$$

From Eqs. (5.15) and (5.16) it is seen that

$$J(x,x,z) = 1 + r(z) \; e^{-2x/z} + \int_0^1 R(z',z,x) \; w(z') \; dz' \; .$$

$$(5.22)$$

The final equation of Riccati type for R is

$$R_x(v,z,x) = - \left(\frac{1}{v} + \frac{1}{z} \right) R(v,z,x)$$

$$+ \left[1 + r(z) \; e^{-2x/z} + \int_0^1 R(z',z,x) \; w(z') \; dz' \right]$$

$$\cdot \left[1 + r(v) \; e^{-2x/v} + \int_0^1 R(v,z',x) \; w(z') \; dz' \right]$$

$$(5.23)$$

From the definition of R in Eq. (5.17) it is seen that

$$R(v,z,0) = 0, \qquad\qquad (5.24)$$

$$0 \leq v, z \leq 1.$$

The differential equation (5.14) for J can now be rewritten in the form

$$J_x(t,x,z) = -\frac{1}{z} J(t,x,z) + \left[\int_0^1 J(t,x,z') \; w(z') \; dz' \right]$$

$$\cdot \left[1 + r(z) \; e^{-2x/z} + \int_0^1 R(z',z,x) \; w(z') \; dz' \right],$$

$$x > t.$$

$$(5.25)$$

The initial condition at x = t is

$$J(t,t,z) = 1 + r(z) \; e^{-2t/z} + \int_0^1 R(z',z,t) \; w(z') \; dz' \; .$$

$$(5.26)$$

Attention is next focused on $u(x,x)$, the second factor on the right-hand side of Eq. (5.8). From the integral equation (5.5) it is noted that

$$u(x,x) = g(x) + \int_0^x k(x,y) \; u(y,x) \; dy \; . \qquad (5.27)$$

This is rewritten as

$$u(x,x) = g(x) + \int_0^x \int_0^1 \left[e^{-(x-y)/z'} \right.$$

$$\left. + r(z') \, e^{-(x+y)/z'} \right] \, w(z') \, dz' \, u(y,x) \, dy \; ,$$

$$(5.28)$$

or

$$u(x,x) = g(x) + \int_0^1 e(z',x) \, w(z') \, dz' \; , \qquad (5.29)$$

where

$$e(z,x) = \int_0^x \left[e^{-(x-y)/z} + r(z) \, e^{-(x+y)/z} \right] u(y,x) \, dy \; .$$

$$(5.30)$$

By differentiating this last equation with respect to x we obtain the relation

$$e_x(z,x) = -\frac{1}{z} \, e(z,x) + \left[1 + r(z) \, e^{-2x/z} \right] u(x,x)$$

$$+ \, u(x,x) \int_0^x \left[e^{-(x-y)/z} + r(z) \, e^{-(x+y)/z} \right]$$

$$\cdot \, \Phi(y,x) \, dy \; . \qquad (5.31)$$

Using Eqs. (5.12) and (5.20), the integral in the last equation is seen to be

$$\left[e^{-(x-y)/z} + r(z) \, e^{-(x+y)/z} \right] \Phi(y,x) \, dy$$

$$= \int_0^1 R(z,z',x) \, w(z') \, dz' \; . \qquad (5.32)$$

The differential equation for the function e becomes

$$e_x(z,x) = -\frac{1}{z} e(z,x) + \left[g(x) + \int_0^1 e(z',x) \; w(z') \; dz'\right]$$

$$\cdot \left[1 + r(z) \; e^{-2x/z} + \int_0^1 R(z,z',x) \; w(z') \; dz'\right] .$$

(5.33)

The initial condition for e at x = 0 is

$$e(z,0) = 0,$$

(5.34)

$$0 \leq z \leq 1.$$

The differential equation for the function u has already been obtained:

$$u_x(t,x) = \left[\int_0^1 J(t,x,z') \; w(z') \; dz'\right]$$

$$\cdot \left[g(x) + \int_0^1 e(z',x) \; w(z') \; dz'\right] ,$$

(5.35)

$$x > t.$$

The initial condition on u at x = t is

$$u(t,t) = g(t) + \int_0^1 e(z',t) \; w(z') \; dz' .$$

(5.36)

5.3   STATEMENT OF THE CAUCHY PROBLEM

The initial value problem for the function u and

the auxiliary functions R, e, and J can now be stated in com
pact form. This provides a convenient summary of the previously derived equations. The functions R and e are determined for $x_1 > x \geq 0$ (when x_1 is sufficiently small)
by the relations

$$R_x(v,z,x) = -\left(\frac{1}{v} + \frac{1}{z}\right) R(v,z,x)$$

$$+ \left[1 + r(z) \ e^{-2x/z} + \int_0^1 R(z',z,x) \ w(z') \ dz'\right]$$

$$\cdot \left[1 + r(v) \ e^{-2x/v} + \int_0^1 R(v,z',x) \ w(z') \ dz'\right]$$

$$x \geq 0,$$

$$R(v,z,0) = 0,$$

$$0 \leq v, z \leq 1,$$

(5.37)

(5.38)

$$e_x(z,x) = -\frac{1}{z} e(z,x) + \left[g(x) + \int_0^1 e(z',x) \ w(z') \ dz'\right]$$

$$\cdot \left[1 + r(z) \ e^{-2x/z} + \int_0^1 R(z,z',x) \ w(z') \ dz'\right],$$

$$x \geq 0,$$

(5.39)

and

$$e(z,0) = 0 \ ,$$

$$0 \leq z \leq 1.$$

(5.40)

At $x = t < x_1$ the functions $J(t,x,z)$ and $u(t,x)$ are then given by

$$u(t,t) = g(t) + \int_0^1 e(z',t) \, w(z') \, dz' \qquad (5.41)$$

$$J(t,t,z) = 1 + r(z) \, e^{-2x/z} + \int_0^1 R(z',z,t) \, w(z') \, dz' \, ,$$
$$0 \le z \le 1.$$
$$(5.42)$$

The differential equations for u and J, valid for $x_1 > x > t$, are

$$u_x(t,x) = \left[\int_0^1 J(t,x,z') \, w(z') \, dz' \right]$$

$$\cdot \left[g(x) + \int_0^1 e(z',x) \, w(z') \, dz' \right] , \qquad (5.43)$$

and

$$J_x(t,x,z) = -\frac{1}{z} J(t,x,z) + \left[\int_0^1 J(t,x,z') \, w(z') \, dz' \right]$$

$$\cdot \left[1 + r(z) \, e^{-x/z} + \int_0^1 R(z',z,x) \, w(z') \, dz' \right] .$$
$$(5.44)$$

For the case $r \equiv 0$, that is, when $k_2(t+y) \equiv 0$, these equations reduce to the known equations of Chapter 4. The case $r \equiv 1$ is especially interesting for certain applications.

5.4   NUMERICAL METHOD

The numerical method consists in approximating integrals on the right-hand sides of Eqs. (5.37), (5.39), (5.43), and (5.44) by the use of Gaussian quadrature finite sums. Let the integral   I,

$$I = \int_0^1 f(z') \, w(z') \, dz' \, , \tag{5.45}$$

be approximated by the formula

$$I \cong \sum_{i=1}^{N} f(\alpha_i) \, \beta_i \, . \tag{5.46}$$

Then Eq. (5.37) becomes the Riccati system

$$R'_{ij} = - \left(\frac{1}{\alpha_i} + \frac{1}{\alpha_j} \right) R_{ij}$$

$$+ \left[1 + r(\alpha_j) \, e^{-2x/\alpha_j} + \sum_{m=1}^{N} R_{mj} \, \beta_m \right] \tag{5.47}$$

$$\cdot \left[1 + r(\alpha_i) \, e^{-2x/\alpha_i} + \sum_{m=1}^{N} R_{im} \, \beta_m \right] \, ,$$

where the prime on R_{ij} indicates differentiation with respect to   x.   This system of N^2 equations is readily integrated numerically, when $x < x_1$,   using the initial conditions

$$R_{ij}(0) = 0, \tag{5.48}$$
$$i,j=1,2,\ldots,N.$$

The symmetry of R_{ij},

$$R_{ij}(x) = R_{ji}(x) , \qquad (5.49)$$

may be used for numerical checking purposes or to reduce the number of equations from N^2 to $N(N+1)/2$. Previous experiments on related problems have shown that $N \cong 7$ is appropriate; the quadrature formula is very accurate, and good coverage of the dependence of R on v and z is obtained.

Similar considerations apply to obtaining the values of the functions e, J, and u. Note that we may either calculate R and e first, store them, and then compute J and u for $x > t$, or calculate R and e on the interval $0 \leq x \leq x_1$. The latter procedure is the one used here, since it eliminates problems of limited computer storage.

5.5   DISCUSSION

Using this method, numerical results have been obtained for the solution of Eq. (5.5) where

$$g(t) = \frac{\lambda}{4} \left[e^{-(x-t)/u} + p\, e^{-(x+t)/u} \right], \qquad (5.50)$$

and

$$k_1(s) = \int_0^1 e^{-s/z}\, \frac{dz}{z} ,$$

$$\qquad (5.51)$$

$$k_2(s) = p\, k_1(s) ,$$

$$0 \leq p \leq 1.$$

They are reported in the reference cited below.

It remains to be established that the solution of the initial value problem does indeed satisfy the integral equation. This has been done in the case $r(v) \equiv 0$, and the proof probably carries over to $r \neq 0$. We leave this as an exercise for the reader.

The formalism presented in this chapter opens the way to the treatment of many inverse or system identification problems by the method of quasilinearization. In particular, if the reflection function $R(v,z,x)$ is measured experimentally, it is possible to estimate the specular reflector function $r(v)$.

It is of special interest to study the behavior of the function $u(t,x)$ for large x, and to relate this to the solution of the equation

$$u(t) = g(t) + \int_0^\infty k(t,y) \, u(y) \, dy \, , \tag{5.52}$$

$$0 \leq t < \infty.$$

In some instances "critical lengths" will exist at which the solution of Eqs. (5.37) and (5.38) will become infinite.

Another interesting type of composite kernel is

$$k(t,y) = E_1(|t-y|) + AE_2(t) \, E_2(y) \tag{5.53}$$

where

$$E_1(r) = \int_0^1 e^{-r/z} \, dz/z \, , \tag{5.54}$$

$$E_2(r) = \int_0^1 e^{-r/z} \, dz \, , \qquad\qquad r > 0. \tag{5.55}$$

In this case, as was done earlier for kernels of purely displacement type, X and Y functions may be substituted for the R function, with the advantage of a reduction in dimensionality. This is also conceptually exciting because of the implication that X and Y functions may occur more widely than is currently recognized. The imbedding approach is used to treat an integral equation of this type appearing in the theory of radiative transfer in Chapter 9.

EXERCISES

1. Validate the initial value problem which is derived in this chapter.

2. Write a computer program and reproduce some of the results given in the reference.

REFERENCE

J. Casti, R. Kalaba, and S. Ueno, "Source Functions for an Isotropically Scattering Atmosphere Bounded by a Specular Reflector," *Journal of Quant. Spectrosc. Radiat. Transfer*, Vol. 10 (1970), pp. 1119-1128.

CHAPTER 6

FREDHOLM RESOLVENTS FOR GENERAL KERNELS

6.1 INTRODUCTION

In previous chapters we derived Cauchy problems for
linear Fredholm integral equations under various assumptions
on the form of the kernel. Here we will derive for *general*
kernels an initial value problem in which the independent
variable is the interval length, assuming merely that certain
differentiations and integrations are valid.

Consider the Fredholm integral equation

$$u(t) = g(t) + \int_0^x k(t,y')\, u(y')\, dy' , \qquad (6.1)$$

$$0 \leq t \leq x,$$

and assume that a unique solution exists for all interval
lengths x for which $0 \leq x \leq x_1$. In terms of the resolvent
K, the solution of this equation is

$$u(t) = g(t) + \int_0^x K(t,y',x)\, g(y')\, dy' . \qquad (6.2)$$

215

We assume that both the kernel k and the resolvent K
are bounded, continuous, and differentiable. The resolvent
kernel itself satisfies the Fredholm integral equation

$$K(t,y,x) = k(t,y) + \int_0^x k(t,y') \ K(y',y,x) \ dy', \quad (6.3)$$

$$0 \le t, y \le x.$$

6.2   DERIVATION OF THE CAUCHY SYSTEM

Let us first introduce the normalized independent
variables τ, η, and η' by means of the definitions

$$t = \tau x \ ,$$

$$y = \eta x \ ,$$
$$\hspace{6cm} (6.4)$$
$$y' = \eta' x \ , \hspace{3cm} 0 \le \tau, \eta, \eta' \le 1 \ .$$

Then Eqs. (6.1) and (6.2) are written

$$u(\tau x) = g(\tau x) + x \int_0^1 k(\tau x, \eta' x) \ u(\eta' x) \ d\eta' \ , \quad (6.5)$$

$$0 \le \tau \le 1 \ ,$$

and

$$u(\tau x) = g(\tau x) + x \int_0^1 K(\tau x, \eta' x, x) \ g(\eta' x) \ d\eta' \ . \quad (6.6)$$

Let us introduce the new dependent variable R through the definition

$$R(\tau,\eta,x) = K(\tau x,\eta x,x) . \tag{6.7}$$

In terms of the function R , the solution of the integral equation, Eq. (6.1) or (6.5), is

$$u(\tau x) = g(\tau x) + x \int_0^1 R(\tau,\eta',x)\ g(\eta'x)\ d\eta' , \tag{6.8}$$

because of Eqs. (6.6) and (6.7). In view of Eqs. (6.3) and (6.7), it is clear that R satisfies the integral equation

$$R(\tau,\eta,x) = k(\tau x,\eta x) + x \int_0^1 k(\tau x,\eta'x)\ R(\eta',\eta,x)\ d\eta'. \tag{6.9}$$

Now differentiate both sides of the last equation with respect to x . The result is

$$R_x(\tau,\eta,x) = \frac{d}{dx}\ k(\tau x,\eta x)$$

$$+ \int_0^1 k(\tau x,\eta'x)\ R(\eta',\eta,x)\ d\eta'$$

$$\tag{6.10}$$

$$+ x \int_0^1 \frac{d}{dx}\ k(\tau x,\eta'x)\ R(\eta',\eta,x)\ d\eta'$$

$$+ x \int_0^1 k(\tau x,\eta'x)\ R_x(\eta',\eta,x)\ d\eta' .$$

As usual, the subscript denotes partial differentiation.
Let Q denote the first three terms on the right-hand side
of Eq. (6.10), namely

$$Q(\tau,\eta,x) = \frac{d}{dx} k(\tau x, \eta x)$$

$$+ \int_0^1 k(\tau x, \eta' x) \, R(\eta',\eta,x) \, d\eta' \qquad (6.11)$$

$$+ x \int_0^1 \frac{d}{dx} k(\tau x, \eta' x) \, R(\eta',\eta,x) \, d\eta' \ .$$

Then Eq. (6.10) can be written

$$R_x(\tau,\eta,x) = Q(\tau,\eta,x)$$

$$\qquad (6.12)$$

$$+ x \int_0^1 k(\tau x, \eta' x) \, R_x(\eta',\eta,x) \, d\eta' \ .$$

We regard this as an integral equation for the function R_x.
Since this equation has the form of Eq. (6.5), its solution can be expressed as

$$R_x(\tau,\eta,x) = Q(\tau,\eta,x)$$

$$\qquad (6.13)$$

$$+ x \int_0^1 R(\tau,\eta',x) \, Q(\eta',\eta,x) \, d\eta' \ .$$

The total derivative of the kernel is written in terms of
the partial derivatives k_1 and k_2 where

$$k_1(t,y) = \partial k/\partial t ,$$

(6.14)

$$k_2(t,y) = \partial k/\partial y ,$$

and

$$\frac{d}{dx} k(\tau x,\eta x) = \tau k_1(\tau x,\eta x)$$

(6.15)

$$+ \eta k_2(\tau x,\eta x) ,$$

so that

$$Q(\tau,\eta,x) = \tau k_1(\tau x,\eta x) + \eta k_2(\tau x,\eta x)$$

$$+ \int_0^1 [k(\tau x,\eta' x) + x\tau k_1(\tau x,\eta' x)$$

(6.16)

$$+ x\eta' k_2(\tau x,\eta' x)] R(\eta',\eta,x) d\eta' .$$

The initial condition on R is

$$R(\tau,\eta,0) = k(0,0) ,$$

(6.17)

$$0 \le \tau,\eta \le 1,$$

which follows from setting $x = 0$ in Eq. (6.9).

The Cauchy system for the resolvent R consists of the differential equation in Eq. (6.13), the definition of the auxiliary variable Q in Eq. (6.16), and the initial condition in Eq. (6.17).

6.3   VALIDATION

Next let us show that a solution of the Cauchy system just described is a solution of the integral equation (6.9). Introduce the function L to be

$$L(\tau,\eta,x) = k(\tau x,\eta x) + x \int_0^1 k(\tau x,\eta'x)\ R(\eta',\eta,x)\ d\eta'\ ,$$

$$0 \le \tau,\eta \le 1,\quad 0 \le x \le x_1,$$

$$(6.18)$$

where R is the solution of the Cauchy system in Eqs. (6.13), (6.16), and (6.17). We plan to show that $L \equiv R$. Differentiation with respect to x shows that

$$L_x(\tau,\eta,x) = \tau k_1(\tau x,\eta x) + \eta k_2(\tau x,\eta x)$$

$$+ \int_0^1 k(\tau x,\eta'x)\ R(\eta',\eta,x)\ d\eta' \qquad (6.19)$$

$$+ x \int_0^1 [\tau k_1(\tau x,\eta'x) + \eta' k_2(\tau x,\eta'x)] R(\eta',\eta,x)\ d$$

$$+ x \int_0^1 k(\tau x,\eta'x)\ R_x(\eta',\eta,x)\ d\eta'$$

$$= Q(\tau,\eta,x) + x \int_0^1 k(\tau x,\eta'x)\ R_x(\eta',\eta,x)\ d\eta'\ .$$

Making use of the differential equation (6.13) for the resolvent R we may rewrite the above equation in the form

$$L_x(\tau,\eta,x) = Q(\tau,\eta,x) + x \int_0^1 k(\tau x,\eta'x) \Big[Q(\eta',\eta,x)$$

$$(6.20)$$

$$+ x \int_0^1 R(\eta',\eta'',x) \; Q(\eta'',\eta,x) \; d\eta'' \Big] \; d\eta' \; .$$

In turn this equation may be written in the form

$$L_x(\tau,\eta,x) = Q(\tau,\eta,x) + x \int_0^1 \Big[k(\tau x,\eta'x)$$

$$+ x \int_0^1 k(\tau x,\eta''x) \; R(\eta'',\eta',x) \; d\eta'' \Big] \qquad (6.21)$$

$$\cdot \; Q(\eta',\eta,x) \; d\eta' \; .$$

From the definition of the function L in Eq. (6.18), L
satisfies the differential equation

$$L_x(\tau,\eta,x) = Q(\tau,\eta,x) + x \int_0^1 L(\tau,\eta',x) \; Q(\eta',\eta,x) \; d\eta' \; .$$

$$(6.22)$$

It also satisfies the initial condition

$$L(\tau,\eta,0) = k(0,0) \; , \qquad\qquad\qquad (6.23)$$

$$0 \le \tau,\eta \le 1.$$

Comparing Eqs. (6.22) and (6.23) for the function L against
Eqs. (6.13) and (6.17) for R and assuming uniqueness of
solution, we find that,

$$R(\tau,\eta,x) = L(\tau,\eta,x) \qquad\qquad\qquad (6.24)$$

or

$$R(\tau,\eta,x) = k(\tau,\eta x) + x\int_0^1 k(\tau x,\eta'x)\ R(\eta',\eta,x)\ d\eta'\ ,$$

$$(6.25)$$

$$0\le\tau,\eta\le 1,\quad 0\le x\le x_1,$$

which was to be proved.

We have now established that the Cauchy system in Eqs. (6.13), (6.16) and (6.17) is equivalent to the integral equation in Eq. (6.9)

6.4   COMPUTATIONAL ASPECTS

Much previous computational experience with related Cauchy systems leads us to believe that the method of lines will be effective numerically. The basic idea is to approximate integrals on the interval [0,1] by a quadrature formula of the form

$$\int_0^1 f(y')\ dy' \cong \sum_{j=1}^N f(r_j)\ w_j\ .$$

$$(6.26)$$

Then we introduce the nomenclature

$$R_{ij}(x) = R(r_i,r_j,x)\ ,$$

$$(6.27)$$

$$i,j=1,2,\ldots,N,\quad 0\le x\le x_1,$$

and

$$Q_{ij}(x) = Q(r_i, r_j, x) , \tag{6.28}$$

$$i,j=1,2,\ldots,N, \quad 0\leq x\leq x_1 ,$$

The exact equations (6.13) and (6.16) are replaced by the approximations

$$R_{ij}'(x) = Q_{ij}(x) + x \sum_{m=1}^{N} R_{im}(x) \, Q_{mj}(x) \, w_m , \tag{6.29}$$

and

$$Q_{ij}(x) = r_i k_1(r_i x, r_j x) + r_j k_2(r_i x, r_j x)$$

$$+ \sum_{m=1}^{N} [k(r_i x, r_m x) + xr_i k_1(r_i x, r_m x) \tag{6.30}$$

$$+ xr_m k_2(r_i x, r_m x)] \, R_{mj}(x) \, w_m ,$$

$$i,j=1,2,\ldots,N, \quad 0\leq x\leq x_1 .$$

The initial conditions on R_{ij} at $x = 0$ are

$$R_{ij}(0) = k(0,0) , \tag{6.31}$$

$$i,j=1,2,\ldots,N,$$

according to Eq. (6.17).

At the expense of integrating a system of N^2 ordinary differential equations with known initial conditions we obtain the resolvent R when the first two arguments are integration abscissas. But by means of a few summations we may obtain values of R for arbitrary values of the first two arguments, no further integrations being required.

First we recall that R also satisfies the integral equation

$$R(\tau,\eta,x) = k(\tau x,\eta x) + x \int_0^1 R(\tau,\eta',x)\, k(\eta'x,\eta x)\, d\eta' \; .$$

$$(6.32)$$

Approximately, we have, then,

$$R(\tau,\eta,x) \cong k(\tau x,\eta x) + x \sum_{n=1}^N R(\tau,r_n,x)\, k(r_n x,\eta x)\, w_n \; .$$

$$(6.33)$$

In particular for $\tau = r_1, r_2, \ldots, r_N$ we have

$$R(r_i,\eta,x) \cong k(r_i x,\eta x) + x \sum_{n=1}^N R_{in}(x)\, k(r_n x,\eta x)\, w_n \; .$$

$$(6.34)$$

This permits us to evaluate R when the second argument is arbitrary, but the first is a quadrature abscissa.   Next we recall Eq. (6.9) and write it in the approximate form

$$R(\tau,\eta,x) \cong k(\tau x,\eta x) + x \sum_{n=1}^N k(\tau x,r_n x)\, R(r_n,\eta,x)\, w_n \; .$$

$$(6.35)$$

This is especially useful if we wish to change the order of the quadrature formula or do numerical checks as the integration of Eq. (6.13) proceeds.

   With regard to the solution of the integral equation (6.5) itself, we use Eq. (6.8) and observe that

$$u(r_m x) \cong g(r_m x) + x \sum_{n=1}^{N} R_{mn}(x) \, g(r_n x) \, w_n \, , \qquad (6.36)$$

$$m = 1, 2, \ldots, N.$$

Once the quantities $u(r_1 x)$, $u(r_2 x), \ldots, u(r_N x)$ have been evaluated, we may return to Eq. (6.5) and write, for arbitrary τ,

$$u(\tau x) \cong g(\tau x) + x \sum_{n=1}^{N} k(\tau x, r_n x) \, u(r_n x) \, w_n \, . \qquad (6.37)$$

Thus $u(t)$, for arbitrary t, may be evaluated from a knowledge of $R_{mn}(x)$, no further integrations being required.

6.5   NUMERICAL RESULTS

Numerical experiments are undertaken to show the feasibility, practicality, and accuracy of the method. For these examples, Simpson's rule with 7 points is used to approximate the integrals. The systems of ordinary differential equations are numerically integrated using a fourth-order Runge-Kutta method with step size $\Delta x = 0.05$. The calculations are performed on a General Electric 265 timesharing computer.

In Example 1, the kernel is

$$k(t,y) = t \, y \, . \qquad (6.38)$$

The analytical solution for the resolvent is

$$K(t,y,x) = t \, y \left[1 - \frac{x^3}{3} \right]^{-1} \, . \qquad (6.39)$$

The resolvent is calculated by the initial value method, and six-digit accuracy is found, as shown in Table 6.1. (Computational results obtained by K. Spingarn.)

Table 6.1

Resolvent R for kernel k(t,y) = ty
N=7, Δx=1/20, x=1

Initial value method

τ \ η	0	1/6	1/3	1/2	2/3	5/6	1
0	0	0	0	0	0	0	0
1/6	0	.0416667	.0833334	.125	.166667	.208333	.25
1/3	0	.0833334	.166667	.25	.333333	.416667	.5
1/2	0	.125	.25	.375	.5	.625	.75
2/3	0	.166667	.333333	.5	.666667	.833334	1.
5/6	0	.208333	.416667	.625	.833334	1.04167	1.25
1	0	.25	.5	.75	1.	1.25	1.5

Analytical solution

τ \ η	0	1/6	1/3	1/2	2/3	5/6	1
0	0	0	0	0	0	0	0
1/6	0	.0416667	.0833333	.125	.166667	.208333	.25
1/3	0	.0833333	.166667	.25	.333333	.416667	.5
1/2	0	.125	.25	.375	.5	.625	.75
2/3	0	.166667	.333333	.5	.666667	.833333	1.
5/6	0	.208333	.416667	.625	.833333	1.04167	1.25
1	0	.25	.5	.75	1.	1.25	1.5

When the forcing function is $g(t) = 1$, the solution
of the integral equation is

$$u(t) = 1 + x^2 t \left[2(1 - \frac{x^3}{3}) \right]^{-1} . \tag{6.40}$$

The high accuracy of the imbedding method is shown in Table
6.2.

Table 6.2

Solution $u(t)$ for kernel $k(t,y) = ty$

t	Initial value method	Analytical solution
0	1.	1.
1/6	1.125	1.125
1/3	1.25	1.25
1/2	1.375	1.375
2/3	1.5	1.5
5/6	1.625	1.625
1	1.75	1.75

Example 2 has the kernel

$$k(t,y) = P \, e^{-\alpha |t-y|} \tag{6.41}$$

This type of kernel appears in communication theory as the
covariance of a stationary random process with a rational
spectrum. The derivatives of the kernel are discontinuous
when $t = y$; they are arbitrarily defined such that

$$k_1(t,y) = \begin{cases} -\alpha\,P\,e^{-(t-y)}, & t \geq y, \\ \\ +\alpha\,P\,e^{-(y-t)}, & t < y, \end{cases} \qquad (6.42)$$

and

$$k_2(t,y) = \begin{cases} +\alpha\,P\,e^{-(t-y)}, & t \geq y, \\ \\ -\alpha\,P\,e^{-(y-t)}, & t < y. \end{cases} \qquad (6.43)$$

When $P = 1/2$, $\alpha = 1$, and $x = 1$, the resolvent is

$$K(t,y,x) = \begin{cases} (2-t)(1+y)/3, & t \geq y, \\ \\ (2-y)(1+t)/3, & t \leq y. \end{cases} \qquad (6.44)$$

Results of the experiment are shown in Table 6.3. Up to three-digit accuracy is obtained, and the maximum error is less than 1.2 percent. The increased error over that of the first example is believed to be caused by the discontinuity in the derivatives. Additional results not shown in the table show that increasing the step size to $\Delta x = 0.20$ has no effect on accuracy. However, increasing the number of points in Simpson's rule from 7 to 9 and using $\Delta x = 0.10$ led to a maximum error of only 0.6 percent or less.

The solution when $g(t) = 1$ is found to be

$$u(t) = (3 + t - t^2)/2. \qquad (6.45)$$

Table 6.3

Resolvent R for kernel $k(t,y) = (1/2)\, e^{-|t-y|}$

Initial-value method

τ \ η	0	1/6	1/3	1/2	2/3	5/6	1
0	.66747	.614874	.556519	.503305	.445377	.391563	.334082
1/6	.614874	.720668	.652273	.589903	.522008	.458935	.391563
1/3	.556519	.652273	.741917	.670976	.593749	.522008	.445377
1/2	.503305	.589903	.670976	.758247	.670976	.589903	.503305
2/3	.445377	.522008	.593749	.670976	.741917	.652273	.556519
5/6	.391563	.458935	.522008	.589903	.652273	.720668	.614874
1	.334082	.391563	.445377	.503305	.556519	.614874	.66747

Analytical solution

τ \ η	0	1/6	1/3	1/2	2/3	5/6	1
0	.666667	.611111	.555556	.5	.444444	.388889	.333333
1/6	.611111	.712963	.648148	.583333	.518519	.453704	.388889
1/3	.555556	.648148	.740741	.666667	.592593	.518519	.444444
1/2	.5	.583333	.666667	.75	.666667	.583333	.5
2/3	.444444	.518519	.592593	.666667	.740741	.648148	.555556
5/6	.388889	.453704	.518519	.583333	.648148	.712963	.611111
1	.333333	.388889	.444444	.5	.555556	.611111	.666667

Results are presented in Table 6.4. The maximum
error is less than 0.66 percent. This can be reduced to
less than 0.37 percent by increasing the number of quadrature points from 7 to 9 and decreasing Δx to 0.10.

These experiments show how well the method works,
even for kernels with discontinuous derivatives. Further
study is required to determine the most efficient method
of implementation.

Table 6.4

Solution u(t) for kernel $k(t,y) = (1/2)\ e^{-|t-y|}$

t	Initial value method	Analytical solution
0	1.50246	1.5
1/6	1.57961	1.56944
1/3	1.61413	1.61111
1/2	1.63571	1.625
2/3	1.61413	1.61111
5/6	1.57961	1.56944
1	1.50246	1.5

6.6   DISCUSSION

In this chapter we have presented a Cauchy system for
Fredholm resolvents for general kernels and have shown that
it is both necessary and sufficient. Some of the numerical
aspects have been discussed, and results of experiments have
been presented. The breadth of applicability is greatly

extended by recalling that various techniques are available for smoothing singular kernels.

In the next chapter, we shall also discuss nonlinear integral equations from this point of view.

EXERCISES

1. Write a computer program for calculating resolvents of integral operators with general kernels.

2. Use the program to produce solutions for the three examples in this chapter. Compare results.

REFERENCE

R. Kalaba and K. Spingarn, "Numerical Solution of the Integral Integral Equations for Optimal Sequential Filtering," *Proc. Hawaii Conference.*

CHAPTER 7

NONLINEAR INTEGRAL EQUATIONS

7.1   INTRODUCTION

Earlier chapters have been confined to linear integral equations. However, we are forced in such areas as optimal filtering and lateral inhibition in neural systems to
deal with nonlinear integral equations of the type

$$u(t) = g(t) + \int_0^x k(t,y,u(y))\ dy\ ,$$

$$0 \leq t \leq x.$$

This nonlinear integral equation may be converted to an initial-
value problem, as we shall see in this chapter. We make no
approximations or expansions; we require only certain differ-
entiability properties.

The complete system of differential equations is obtained through differentiation with respect to x, the upper-
limit, and realization that the derivative of u satisfies
a linear integral equation. This implies that it would be

useful to introduce the resolvent of the related linear integral equation. The system is closed by deriving an equation for the resolvent.

When the system is specialized in the case of a linear kernel, the equations are in agreement with those in
Chapter 6.

7.2   DERIVATION OF THE CAUCHY SYSTEM

Consider the nonlinear integral equation

$$u(t,x) = g(t) + \int_0^x k(t,y,u(y,x))\, dy , \qquad (7.1)$$

$$0 \le t \le x,$$

and assume that it has a differentiable solution for all
interval lengths x such that $0 \le x \le x_1$. Introduce the
variables τ and η via the definitions

$$\tau = t/x, \qquad\qquad\qquad 0 \le \tau \le 1 , \qquad (7.2)$$

and

$$\eta = y/x , \qquad\qquad\qquad 0 \le \eta \le 1 . \qquad (7.3)$$

Eq. (7.1) becomes

$$u(\tau x,x) = g(\tau x) + x \int_0^1 k(\tau x,\eta x,\ u(\eta x,x))\, d\eta . \qquad (7.4)$$

$$0 \le \tau \le 1, \quad 0 \le x \le x_1$$

Also introduce the new dependent variable v   according to
the definition

$$v(\tau,x) = u(\tau x, x),\qquad\qquad\qquad (7.5)$$
$$0 \le \tau \le 1, \quad 0 \le x \le x_1,$$

so that the integral equation (7.4) becomes

$$v(\tau,x) = g(\tau x) + x \int_0^1 k(\tau x, \eta x, v(\eta,x))\,d\eta,\qquad (7.6)$$
$$0 \le \tau \le 1, \quad 0 \le x \le x_1.$$

Now differentiate both sides of Eq. (7.6) with respect to
x .   The result is

$$v_x(\tau,x) = F(\tau,x)$$

$$+ x \int_0^1 k_3(\tau x, \eta x, v(\eta,x)) v_x(\eta,x)\,d\eta.\qquad (7.7)$$

where

$$F(\tau,x) = \tau g'(\tau x) + \int_0^1 k(\tau x, \eta x, v(\eta,x))\,d\eta$$

$$+ x \int_0^1 \tau k_1(\tau x, \eta x, v(\eta,x))\,d\eta\qquad\qquad (7.8)$$

$$+ x \int_0^1 \eta k_2(\tau x, \eta x, v(\eta,x))\,d\eta.$$

In Eqs. (7.7) and (7.8) we have indicated the partial derivative of k with respect to its ith argument by $k_i (i=1,2,3)$,
the derivative of g   is written

$$g'(t) = dg/dt , \tag{7.9}$$

and the subscript x denotes partial differentiation with respect to x .

Eq. (7.7) is viewed as a linear Fredholm integral equation for the function v_x . Note that a partial derivative of the original nonlinear function k is the kernel of the linear integral equation. To solve Eq. (7.7) introduce the resolvent kernel R , which is the solution of the linear integral equation

$$R(\tau,\xi,x) = k_3(\tau x,\xi x,v(\xi,x))$$

$$+ x \int_0^1 k_3(\tau x,\eta x,v(\eta,x)) R(\eta,\xi,x) \, d\eta, \tag{7.10}$$

$$0\leq\tau,\xi\leq1, \quad 0\leq x < x_1 \; .$$

In terms of the resolvent R , the solution of Eq. (7.7) may be represented as

$$v_x(\tau,x) = F(\tau,x) + x \int_0^1 R(\tau,\eta,x) \, F(\eta,x) \, d\eta \; . \tag{7.11}$$

Let us write this in the form

$$v_x(\tau,x) = G(\tau,x) , \tag{7.12}$$

where

$$G(\tau,x) = F(\tau,x) + x \int_0^1 R(\tau,\eta,x) \, F(\eta,x) \, d\eta \; . \tag{7.13}$$

Eq. (7.12), supplemented by Eqs. (7.13) and (7.8), is one
of the basic differential equations. Since it involves the
unknown function R , by virtue of Eq. (7.13), we shall now
obtain a differential equation for this function to round
out the Cauchy system.

Assuming, for definiteness, that the function k is
twice differentiable, we introduce the auxiliary function
N to be

$$N(\tau,\xi,x) = \frac{d}{dx} k_3(\tau x,\xi x,v(\xi,x))$$

$$= \tau k_{31}(\tau x,\xi x,v(\xi,x)) + \xi k_{32}(\tau x,\xi x,v(\xi,x)) \qquad (7.14)$$

$$+ v_x(\xi,x) k_{33}(\tau x,\xi x,v(\xi,x)) ,$$

using an obvious subscript notation. According to Eq. (7.12)
this becomes

$$N(\tau,\xi,x) = \tau k_{31}(\tau x,\xi x,v(\xi,x)) + \xi k_{32}(\tau x,\xi x,v(\xi,x))$$

$$\qquad (7.15)$$

$$+ G(\xi,x) k_{33}(\tau x,\xi x,v(\xi,x)) .$$

Then return to Eq. (7.10), and through differentiation of
both sides with respect to x, we see that

$$R_x(\tau,\xi,x) = N(\tau,\xi,x) + \int_0^1 k_3(\tau x,\eta x,v(\eta,x))R(\eta,\xi,x) \, d\eta$$

$$\qquad (7.16)$$

$$+ x \int_0^1 N(\tau,\eta,x) R(\eta,\xi,x) \, d\eta$$

$$+ x \int_0^1 k_3(\tau x,\eta x,v(\eta,x)) R_x(\eta,\xi,x) \, d\eta .$$

Eq. (7.16) is a linear integral equation for the function R_x . Let us introduce the function Q as the sum of the three inhomogeneous terms on the right hand side of Eq. (7.16).

$$Q(\tau,\xi,x) = N(\tau,\xi,x) + \int_0^1 [k_3(\tau x, \eta x, v(\eta,x))$$

$$+ xN(\tau,\eta,x)] R(\eta,\xi,x) \, d\eta \, ,$$

$$0 \leq \tau, \xi \leq 1, \quad 0 \leq x \leq x_1 . \tag{7.17}$$

Then the solution of Eq. (7.16), R_x , may be written as

$$R_x(\tau,\xi,x) = Q(\tau,\xi,x) + x \int_0^1 R(\tau,\eta,x) \, Q(\eta,\xi,x) \, d\eta \, ,$$

$$0 \leq \tau, \xi \leq 1 \, , \tag{7.18}$$

which is the desired differential equation for the resolvent R .

The initial conditions at $x = 0$ on the functions v and R are easily seen to be (by Eqs. (7.6) and (7.10))

$$v(\tau,0) = g(0), \qquad\qquad\qquad 0 \leq \tau \leq 1 \, , \tag{7.19}$$

and

$$R(\tau,\xi,0) = k_3(0,0,g(0)) \, , \tag{7.20}$$

$$0 \leq \tau, \xi \leq 1 \, .$$

We can now summarize the Cauchy system for the functions v and R .

7.3   SUMMARY OF CAUCHY SYSTEM

The function   v   and the auxiliary function   K
satisfy the coupled differential equations

$$v_x(\tau,x) = G(\tau,x) \; , \tag{7.21}$$

$$R_x(\tau,\xi,x) = Q(\tau,\xi,x) + x \int_0^1 R(\tau,\eta,x)Q(\eta,\xi,x) \; d\eta \; ,$$

$$0 \le x \le x_1 \; , \qquad 0 \le \tau, \xi \le 1 \; , \tag{7.22}$$

where

$$G(\tau,x) = F(\tau,x) + x \int_0^1 R(\tau,\eta,x) \; F(\eta,x) \; d\eta \; , \tag{7.23}$$

$$F(\tau,x) = \tau g'(\tau x) + \int_0^1 k(\tau x,\eta x,v(\eta,x)) \; d\eta$$

$$+ x \int_0^1 \tau k_1(\tau x,\eta x,v(\eta,x)) \; d\eta \tag{7.24}$$

$$+ x \int_0^1 \eta k_2(\tau x,\eta x,v(\eta,x)) \; d\eta \; ,$$

$$Q(\tau,\xi,x) = N(\tau,\xi,x) + \int_0^1 [k_3(\tau x,\eta x,v(\eta,x))$$

$$\tag{7.25}$$

$$+ xN(\tau,\eta,x)] \; R(\eta,\xi,x) \; d\eta \; ,$$

and

$$N(\tau,\xi,x) = \tau k_{31}(\tau x,\xi x,v(\xi,x)) + \xi k_{32}(\tau x,\xi x,v(\xi,x))$$
$$+ G(\xi,x)k_{33}(\tau x,\xi x,v(\xi,x)), \qquad (7.26)$$

$$0 \leq x \leq x_1, \qquad 0 \leq \tau,\xi \leq 1 .$$

The initial conditions at $x = 0$ are

$$v(\tau,0) = g(0) , \qquad\qquad\qquad 0 \leq \tau \leq 1 , \qquad (7.27)$$

$$R(\tau,\xi,0) = k_3(0,0,g(0)), \qquad\qquad 0 \leq \tau,\xi \leq 1 . \qquad (7.28)$$

7.4   SPECIALIZATION TO LINEAR INTEGRAL EQUATIONS

In the event that the function k is linear in its third argument, so that Eq. (7.1) or, alternatively, Eq. (7.6), reduces to a linear Fredholm integral equation of the second kind, a great simplication takes place. The differential equations in Eqs. (7.21) and (7.22) become uncoupled, and we may solve for the functions v and R separately.

The linear integral equation is

$$u(t,x) = g(t) + \int_0^x k(t,y) \; u(y,x) \; dy , \qquad (7.29)$$

$$0 \leq t \leq x .$$

Let us write

$$k(t,y,u) = k(t,y)u , \qquad\qquad\qquad\qquad (7.30)$$

so that

$$k_3(t,y,u) = k(t,y) \tag{7.31}$$

and

$$k_{33}(t,y,u) = 0 . \tag{7.32}$$

Also we see that

$$k_{31}(t,y,u) = k_1(t,y) \tag{7.33}$$

and

$$k_{32}(t,y,u) = k_2(t,y) . \tag{7.34}$$

It follows that

$$N(\tau,\xi,x) = \tau k_1(\tau x,\xi x) + \xi k_2(\tau x,\xi x) , \tag{7.35}$$

which is independent of the solution v . For the function Q we have

$$
\begin{aligned}
Q(\tau,\xi,x) = &\ \tau k_1(\tau x,\xi x) + \xi k_2(\tau x,\xi x) \\
&+ \int_0^1 [k(\tau x,\eta x) + x\tau k_1(\tau x,\eta x) \\
&+ x\xi k_2(\tau x,\eta x)] \, R(\eta,\xi,x) \, d\eta ,
\end{aligned}
\tag{7.36}
$$

and the function R satisfies the differential equation

$$R_x(\tau,\xi,x) = Q(\tau,\xi,x) + x \int_0^1 R(\tau,\eta,x) \, Q(\eta,\xi,x) \, d\eta \, ,$$

$$0 \le x \le x_1 \, , \qquad 0 \le \tau, \xi \le 1 \, . \qquad (7.37)$$

The initial condition on the function R at x = 0 is, of course,

$$R(\tau,\xi,0) = k(0,0) \, , \qquad\qquad 0 < \tau, \xi \le 1 \, . \quad (7.38)$$

These equations for R agree with those derived in Chapter 6.

To determine the function v we may proceed in one of two ways. We may use the Cauchy system given in the previous section with the appropriate specializations. Or, we may observe that the function R satisfies the integral equation

$$R(\tau,\xi,x) = k(\tau x,\xi x) + x \int_0^1 k(\tau x,\eta x) \, R(\eta,\xi,x) \, d\eta \, ,$$

$$(7.39)$$

which marks it as the resolvent of the integral equation for the function v ,

$$v(\tau,x) = g(\tau x) + x \int_0^1 k(\tau x,\eta x) \, v(\eta,x) \, d\eta \, . \qquad (7.40)$$

The function v is then given by the representation

$$v(\tau,x) = g(\tau x) + x \int_0^1 R(\tau,\eta,x) \, g(\eta x) \, d\eta \, , \qquad (7.41)$$

$$0 \le \tau \le 1, \quad 0 \le x \le x_1 \, .$$

7.5   NUMERICAL ASPECTS

To solve the Cauchy system numerically we use the method of lines. We begin by approximating integrals on the interval [0,1] with the quadrature formula

$$f(y) \ dy \cong \sum_{m=1}^{M} f(r_m) w_m \ . \tag{7.42}$$

Then we introduce the nomenclature

$$v_i(x) = v(r_i,x) \ ,$$

$$G_i(x) = G(r_i,x) \ ,$$

$$F_i(x) = F(r_i,x) \ , \tag{7.43}$$

$$R_{ij}(x) = R(r_i,r_j,x) \ ,$$

$$Q_{ij}(x) = Q(r_i,r_j,x) \ ,$$

$$N_{ij}(x) = N(r_i,r_j,x) \ ,$$

$$i,j=1,2,\ldots,M, \quad 0 \le x \le x_1 \ .$$

The analogues of Eqs. (7.21) to (7.26) are

$$v_i'(x) = G_i(x) \ , \tag{7.44}$$

$$R_{ij}'(x) = Q_{ij}(x) + x \sum_{m=1}^{M} R_{im}(x) \ Q_{mj}(x) \ w_m \ , \tag{7.45}$$

$$G_i(x) = F_i(x) + x \sum_{m=1}^{M} R_{im}(x) F_m(x) w_m , \qquad (7.46)$$

$$F_i(x) = r_i g'(r_i x) + \sum_{m=1}^{M} k(r,x,r_m x, v_m(x)) w_m$$

$$+ x \sum_{m=1}^{M} r_i k_1 (r_i x, r_m x, v_m(x)) w_m \qquad (7.47)$$

$$+ x \sum_{m=1}^{M} r_m k_2 (r_i x, r_m x, v_m(x)) w_m ,$$

$$Q_{ij}(x) = N_{ij}(x) + \sum_{m=1}^{M} \{ k_3 (r_i x, r_m x, v_m(x))$$

$$+ x N_{im}(x) \} R_{mj}(x) w_m , \qquad (7.48)$$

and

$$N_{ij}(x) = r_i k_{31} (r_i x, r_j x, v_j(x)) + r_j k_{32} (r_i x, r_j x, v_j(x))$$

$$+ G_j(x) k_{33} (r_i x, r_j x, v_j(x)) , \qquad (7.49)$$

$$i,j=1,2,\ldots,M, \quad 0 \le x \le x_1 .$$

This, of course, represents a system of $M^2 + M$ ordinary differential equations, the primes in Eqs. (7.44) and (7.45) representing differentiation with respect to x . This is practical for M less than about fifty. The initial conditions at $x = 0$ are

$$v_i(0) = g(0) , \qquad (7.50)$$

and

$$R_{ij}(0) = k_3(0,0,g(0)), \qquad i,j=1,2,\ldots,M \ . \quad (7.51)$$

Much previous experience leaves little doubt concerning the numerical efficacy of this approach.

The calculation yields the values of $v_1(x)$, $v_2(x)$,..., $v_m(x)$ for $0 \le x \le x_1$. Should we desire the value of $v(\tau,x)$, we may write

$$v(\tau,x) \stackrel{\sim}{=} g(\tau x) + x \sum_{m=1}^{M} k(\tau x, \ r_m x, v_m(x)) \ w_m \ , \quad (7.52)$$

$$0 \le \tau \le 1, \quad 0 \le x \le x_1 \ .$$

7.6   DISCUSSION

For the first time in this book, we consider nonlinear integral equations and their reduction to initial value problems in which the interval length is the independent variable. The approach is a direct extension of that used for linear Fredholm integral equations. We show the feasibility of the numerical method, but much remains to be investigated both computationally and analytically.

Nonlinear integral equations may also be treated by parameter imbedding (Chapter 8).

EXERCISES

1. Validate the initial value problem for nonlinear integral equation which is derived in this chapter.
2. Write a computer program for nonlinear integral equations, and apply to specific examples.
3. Look up bifurcation phenomena in a standard textbook on integral equations.
4. Compare the methods of this chapter and of Chapter 8 for nonlinear integral equations.

REFERENCE

H. Kagiwada and R. Kalaba, "An Initial Value Method for Non-linear Integral Equations," *Journal of Optimization Theory and Applications*, Vol. 12, No. 4 (1973), pp. 329-337.

Chapter 8

PARAMETER IMBEDDING

8.1   INTRODUCTION

Several techniques have already been described for
transforming Fredholm integral equations into Cauchy systems.
Since Cauchy problems lend themselves well to numerical
solution on high speed computers, this reduction provides
attractive new computational approaches to the solution of
integral equations. Furthermore these transformations are
not without analytic interest.

In this chapter we treat both linear and nonlinear
integral equations involving a parameter by a method which
we call "parameter imbedding." Three types of equations are
taken up in this chapter: (1) the linear Fredholm integral
equation with a parameter outside the integral,

$$u(t) = g(t) + \lambda \int_0^1 k(t,y)\ u(y)\ dy\ ,$$

(2) the nonlinear integral equation,

$$u(t) = g(t,\lambda) + \lambda \int_0^1 k(t,y,\lambda,u(y))\ dy\ ,$$

and (3) the intermediate linear case (nonlinear parameter)

$$u(t) = g(t) + \lambda \int_0^1 k(t,y,\lambda)\ u(y)\ dy\ .$$

We describe the numerical method, and we report on successful computations for a contact problem in elasticity.

Eigenvalue problems are also treated in a unique Cauchy approach to the homogeneous equation

$$u(t) = \lambda \int_0^1 k(t,y)\ u(y)\ dy\ .$$

8.2 LINEAR INTEGRAL EQUATIONS WITH MULTIPLICATIVE PARAMETERS

SUMMARY. Consider the family of Fredholm integral equations

$$u(t,\lambda) = g(t) + \lambda \int_0^1 k(t,y)\ u(y,\lambda)\ dy\ , \qquad (8.1)$$
$$0 \le t \le 1\ ,$$

for λ having a range of positive definite values near zero. For λ sufficiently near zero it is assumed that there is a unique solution of Eq. (8.1) for all inhomogeneous functions g .

Also consider the Cauchy system for the functions u
and K ,

$$u_\lambda(t,y,\lambda) = \int_0^1 K(t,y,\lambda) \; u(y,\lambda) \; dy, \qquad (8.2)$$

$$K_\lambda(t,y,\lambda) = \int_0^1 K(t,y',\lambda) \; K(y',y,\lambda) \; dy' \; , \qquad (8.3)$$
$$0 \leq t, y \leq 1, \quad \lambda < 0;$$

$$u(t,0) = g(t) \; , \qquad (8.4)$$

$$K(t,y,0) = k(t,y) \; , \qquad\qquad 0 \leq t, y \leq 1. \quad (8.5)$$

We assume that this system possesses a unique solution for
λ sufficiently near zero.

Our aim is to show that the family of Fredholm integral equations and the Cauchy system are equivalent.

DERIVATION. Let us first show that if the function u
satisfies Eq. (8.1), then it also satisfies the Cauchy system.
In terms of the resolvent kernel K the solution of Eq. (8.1)
may be represented in the form

$$u(t,\lambda) = g(t) + \lambda \int_0^1 K(t,y,\lambda) \; g(y) \; dy \; , \qquad (8.6)$$
$$0 \leq t \leq 1, \quad \lambda \geq 0 \; .$$

The resolvent kernel K itself satisfies the integral
equation

$$K(t,y,\lambda) = k(t,y) + \lambda \int_0^1 k(t,y') \ K(y',y,\lambda) \ dy',$$

$$0 \leq t, y \leq 1 \ . \tag{8.7}$$

Differentiate both sides of Eq. (8.1) with respect to λ . The result is

$$u_\lambda(t,\lambda) = \int_0^1 k(t,y) \ u(y,\lambda) \ dy$$

$$+ \lambda \int_0^1 k(t,y) \ u_\lambda(y,\lambda) \ dx. \tag{8.8}$$

Keeping Eq. (8.1) in mind this is rewritten as

$$u_\lambda(t,\lambda) = [u(t,\lambda) - g(t)]\lambda^{-1}$$

$$+ \lambda \int_0^1 k(t,y) \ u_\lambda(y,\lambda) \ dy \ . \tag{8.9}$$

This is a Fredholm integral equation for the function u_λ . According to Eq. (8.6) the solution of this equation is

$$u_\lambda(t,\lambda) = [u(t,\lambda) - g(t)]\lambda^{-1}$$

$$+ \int_0^1 K(t,y,\lambda) \ [u(y,\lambda) - g(y)] \ dy \ , \tag{8.10}$$

or

$$u_\lambda(t,\lambda) = \int_0^1 K(t,y,\lambda) \ u(y,\lambda) \ dy \ . \tag{8.11}$$

This is Eq. (8.2). From Eq. (8.6) the initial condition
at $\lambda = 0$ in Eq. (8.4) follows.

By putting

$$u(y,\lambda) = g(y) + \lambda \int_0^1 K(y,y',\lambda) \; g(y') \; dy' \tag{8.12}$$

in Eq. (8.11) we see that

$$u_\lambda(t,\lambda) = \int_0^1 K(t,y,\lambda) \tag{8.13}$$

$$\cdot \left\{ g(y) + \lambda \int_0^1 K(y,y',\lambda) \; g(y') \; dy' \right\} \; dy \; .$$

Furthermore differentiation of both sides of Eq. (8.6) with
respect to λ shows that

$$u_\lambda(t,\lambda) = \int_0^1 K(t,y,\lambda) \; g(y) \; dy \tag{8.14}$$

$$+ \lambda \int_0^1 K_\lambda(t,y,\lambda) \; g(y) \; dy \; .$$

Comparing Eqs. (8.13) and (8.14) and keeping in mind the
arbitrariness of the function g we conclude that

$$K_\lambda(t,y,\lambda) = \int_0^1 K(t,y',\lambda) \; K(y',y,\lambda) \; dy' \; , \tag{8.15}$$

which is Eq. (8.3). The initial condition in Eq. (8.5)
follows from Eq. (8.7). This completes the proof of the
necessity of the Cauchy system.

VALIDATION. Now assume that the functions K and u are determined uniquely by the Cauchy system in Eqs. (8.2) - (8.5), at least for λ sufficiently near zero. Introduce the function A by means of the definition

$$A(t,y,\lambda) = k(t,y) + \lambda \int_0^1 k(t,y')\ K(y',y,\lambda)\ dy'\ ,$$

$$0 \leq t,y \leq 1,\quad 0 \leq \lambda\ . \qquad (8.16)$$

We shall show that

$$A = K\ . \qquad (8.17)$$

In the first place at $\lambda = 0$ we have

$$A(t,y,0) = k(t,y)\ . \qquad (8.18)$$

Next differentiate both sides of Eq. (8.16) with respect to λ :

$$A_\lambda(t,y,\lambda) = \int_0^1 k(t,y')\ K(t,y',\lambda)\ dy'$$

$$\qquad (8.19)$$

$$+ \lambda \int_0^1 k(t,y')\ K_\lambda(y',y,\lambda)\ dy'\ .$$

According to Eq. (8.3) this becomes

$$A_\lambda(t,y,\lambda)$$

$$= \int_0^1 k(t,y')\ K(y',y,\lambda)\ dy'$$

$$+ \lambda \int_0^1 k(t,y') \int_0^1 K(y',y'',\lambda)\ K(y'',y,\lambda)\ dy''\ dy'$$

$$= \int_0^1 \left\{ k(t,y') + \lambda \int_0^1 k(t,y'')\ K(y'',y',\lambda)dy'' \right\} \qquad (8.20)$$

$$\cdot\ K(y',y,\lambda)\ dy'$$

or

$$A_\lambda(t,y,\lambda) = \int_0^1 A(t,y')\ K(y',y,\lambda)\ dy'\ . \qquad (8.21)$$

Assuming that the linear Cauchy system in Eqs. (8.18) and (8.21) possess a unique solution, we see that

$$A(t,y,\lambda) = K(t,y,\lambda)\ , \qquad (8.22)$$

or

$$K(t,y,\lambda) = k(t,y) + \lambda \int_0^1 k(t,y')\ K(y',y,\lambda)\ dy'\ .$$

$$0 \le t,y \le 1,\quad 0 \le \lambda,\quad (8.23)$$

a desired result.

Next let

$$B(t,\lambda) = g(t) + \lambda \int_0^1 k(t,y) \, u(y,\lambda) \, dy \, , \qquad (8.24)$$
$$0 \le t \le 1, \quad 0 \le \lambda \, .$$

Differentiation with respect to λ yields the relations

$$B_\lambda(t,\lambda) = \int_0^1 k(t,y) \, u(y,\lambda) \, dy$$

$$+ \lambda \int_0^1 k(t,y) \, u_\lambda(y,\lambda) \, dy$$

$$\qquad (8.25)$$

$$= \int_0^1 k(t,y) \, u(y,\lambda) \, dy$$

$$+ \lambda \int_0^1 k(t,y) \int_0^1 K(y,y',\lambda) \, u(y',\lambda) \, dy' \, dy \, .$$

Through a change of variables we find that

$$B_\lambda(t,\lambda)$$

$$= \int_0^1 k(t,y) \, u(y,\lambda) \, dy$$

$$+ \lambda \int_0^1 k(t,y') \int_0^1 K(y',y,\lambda) \, u(y,\lambda) \, dy \, dy' \, , \qquad (8.26)$$

$$= \int_0^1 \left\{ k(t,y) + \lambda \int_0^1 k(t,y') \, K(y',y,\lambda) \, dy' \right\} u(y,\lambda) \, dy$$

or

$$B_\lambda(t,\lambda) = \int_0^1 K(t,y,\lambda) \, u(y,\lambda) \, dy \, . \qquad (8.27)$$

Also at $\lambda = 0$ we have the initial condition

$$B(t,0) = g(t) , \qquad\qquad\qquad 0 \le t \le 1 . \qquad (8.28)$$

Assuming uniqueness of solution, it follows that

$$B(t,\lambda) = u(t,\lambda) \qquad\qquad\qquad\qquad (8.29)$$

or

$$u(t,\lambda) = g(t) + \lambda \int_0^1 k(t,y) \, u(y,\lambda) \, dy , \qquad (8.30)$$
$$0 \le t \le 1, \quad 0 \le \lambda .$$

This completes the validation.

Numerical considerations will be given in a later section. First, though, we wish to perform parameter imbedding on a nonlinear integral equation.

8.3   NONLINEAR INTEGRAL EQUATIONS

DERIVATION. We next consider the class of nonlinear integral equations

$$u(t) = g(t,\lambda) + \lambda \int_0^1 k(t,y,\lambda,u(y)) \, dy , \qquad (8.31)$$
$$0 \le t \le 1, \quad 0 \le \lambda \le \Lambda.$$

To call attention to the dependence of a solution u upon λ , as well as upon t , we shall write

$$u = u(t,\lambda) \; , \qquad\qquad\qquad 0 \le t \le 1, \quad 0 \le \lambda \le \Lambda. \qquad (8.32)$$

Equation (8.31) becomes

$$u(t,\lambda) = g(t,\lambda) + \lambda \int_0^1 k(t,y,\lambda,u(y,\lambda)) \; dy \; , \qquad (8.33)$$
$$0 \le t \le 1, \quad 0 \le \lambda \le \Lambda.$$

The parameter λ may occur naturally, or it may be artificially introduced. Assume that a solution u is differentiable in λ, and differentiate both sides of Equation (8.3) to obtain the relation

$$u_\lambda(t,\lambda) = g_\lambda(t,\lambda) + \int_0^1 k(t,y,\lambda,u(y,\lambda)) \; dy$$

$$+ \; \lambda \int_0^1 k_\lambda(t,y,\lambda,u(y,\lambda)) \; dy \qquad\qquad (8.34)$$

$$+ \; \lambda \int_0^1 k_u(t,y,\lambda,u(y,\lambda)) \; u_\lambda(y,\lambda) \; dy \; .$$

The subscripts denote partial differentiation. Equation (8.34) is viewed as a linear Fredholm integral equation in which the kernel is k_u, and the first three terms on the right hand side are forcing terms.

Assume that the linear Fredholm integral equation

$$w(t,\lambda) = F(t,\lambda) + \lambda \int_0^1 k_u(t,y, \; u(y,\lambda)) \; w(y,\lambda) \; dy \; ,$$
$$0 \le t \le 1 \; , \qquad (8.35)$$

possesses a unique solution for $0 \le \lambda \le \Lambda$, where u is a solution of Equation (8.33), and F is arbitrary. In terms of the resolvent kernel K the solution of Equation (8.35) is represented as

$$w(t,\lambda) = F(t,\lambda) + \lambda \int_0^1 K(t,y,\lambda) \, F(y,\lambda) \, dy , \qquad (8.36)$$
$$0 \le t \le 1 .$$

The resolvent kernel K itself satisfies the integral equation

$$K(t,y,\lambda) = k_u(t,y,\lambda,u(t,\lambda))$$
$$\qquad\qquad (8.37)$$
$$+ \lambda \int_0^1 k_u(t,y',\lambda,u(y',\lambda)) \, K(y',y,\lambda) \, dy' ,$$
$$0 \le t, y \le 1, \quad 0 \le \lambda \le \Lambda.$$

Returning to Equation (8.34) we find that the solution u_λ may be represented in the form

$$u_\lambda(t,\lambda) = \Psi(t,\lambda) + \lambda \int_0^1 K(t,y',\lambda) \, \Psi(y',\lambda) \, dy' , \quad (8.38)$$

where the forcing function Ψ is given by

$$\Psi(t,\lambda) = g_\lambda(t,\lambda) + \int_0^1 k(t,y',\lambda,u(y',\lambda)) \, dy'$$
$$\qquad\qquad (8.39)$$
$$+ \lambda \int_0^1 k_\lambda(t,y',\lambda,u(y',\lambda)) \, dy' ,$$
$$0 \le t \le 1, \quad 0 \le \lambda \le \Lambda.$$

Equations (8.38) and (8.39) form a differential equation for the unknown function u , the other unknown function being the resolvent kernel K . From the Equation (8.33) we see that the initial condition on the function u at $\lambda = 0$ is

$$u(t,0) = g(t,0) , \qquad\qquad 0 \le t \le 1 . \qquad (8.40)$$

Next we obtain a differential equation for the resolvent kernel K . Through differentiation Equation (8.37) becomes

$$K_\lambda(t,y,\lambda) = k_{u\lambda}(t,y,\lambda,u(y,\lambda))$$

$$+ k_{uu}(t,y,\lambda,u(y,\lambda)) \; u_\lambda(y,\lambda)$$

$$+ \int_0^1 k_u(t,y',\lambda,u(y',\lambda)) \; K(y',y,\lambda) \; dy'$$

$$(8.41)$$

$$+ \lambda \int_0^1 k_{u\lambda}(t,y',\lambda,u(y',\lambda)) \; K(y',y,\lambda) dy'$$

$$+ \lambda \int_0^1 k_{uu}(t,y',\lambda,u(y',\lambda)) u_\lambda(y',\lambda) K(y',y,\lambda) dy'$$

$$+ \lambda \int_0^1 k_u(t,y',\lambda,u(y',\lambda)) K_\lambda(y',y,\lambda) \; dy' \;.$$

This is considered to be a linear Fredholm integral equation for the function K_λ . The kernel is as in Equation (8.35), and the first five terms on the right hand side are taken to be inhomogeneous forcing terms. For simplicity we intro-duce the auxiliary function Q to be the sum of these

forcing terms, and making use of Equation (8.38) for u ,
we see that

$$Q(t,y,\lambda) = k_{u\lambda}(t,y,\lambda,u(y,\lambda))$$

$$+ k_{uu}(t,y,\lambda,u(y,\lambda))$$

$$\cdot \left\{ \Psi(y,\lambda) + \lambda \int_0^1 K(y,y',\lambda)\ \Psi(y',\lambda)\ dy' \right\}$$

$$+ \int_0^1 k_u(t,y',\lambda,u(y',\lambda))\ K(y',y,\lambda)\ dy'$$

(8.42)

$$+ \lambda \int_0^1 k_{u\lambda}(t,y',\lambda,u(y',\lambda))\ K(y',y,\lambda)\ dy'$$

$$+ \lambda \int_0^1 k_{uu}(t,y',\lambda,u(y',\lambda))$$

$$\cdot \left\{ \Psi(y',\lambda) + \lambda \int_0^1 K(y',y'',\lambda)\ \Psi(y'',\lambda)\ dy'' \right\}$$

$$\cdot\ K(y',y,\lambda)\ dy'\ .$$

Then the integral equation for K_λ is

$$K_\lambda(t,y,\lambda) = Q(t,y,\lambda)$$

(8.43)

$$+ \lambda \int_0^1 k_u(t,y',\ u(y',\lambda))\ K_\lambda(y',y,\lambda)\ dy'\ .$$

It follows that the solution of Equation (8.41) may be
written in the form

$$K_\lambda(t,y,\lambda) = Q(t,y,\lambda) + \lambda \int_0^1 K(t,y',\lambda)\, Q(y',y,\lambda)\, dy' \, ,$$

$$0 \le t, y \le 1, \quad 0 \le \lambda \le \Lambda. \qquad (8.44)$$

The initial condition on the resolvent kernel K at $\lambda = 0$
is

$$K(t,y,0) = k_u(t,y,0,g(y,0)), \qquad\qquad (8.45)$$

$$0 \le t, y \le 1 \, ,$$

which follows from Equation (8.37).

Let us now summarize the Cauchy system for the functions u and K . The initial conditions at $\lambda = 0$ are
given in Equations (8.40) and (8.45). The differential
equations are Equations (8.38) and (8.44), the auxiliary
functions Ψ and Q being defined in Equations (8.39) and
(8.42).

Notice that solution of the Cauchy system provides
the desired parameter study in λ , and no successive
approximations or series expansions are used.

VALIDATION. We shall now show that a solution of the Cauchy
system in Eqs. (8.40), (8.45), (8.38), (8.44), (8.38) and
(8.42) provides a solution of the nonlinear integral equation
Equation (8.33). we begin by showing that K , as determined
by the Cauchy system, satisfies Equation (8.37). Introduce
the auxiliary function A by means of the definition

$$A(t,y,\lambda) = k_u(t,y,\lambda,u(y,\lambda))$$

$$+ \lambda \int_0^1 k_u(t,y',\lambda,u(y',\lambda))\ K(y',y,\lambda)\ dy'\ , \tag{8.46}$$

$$0 \leq t, y \leq 1, \quad 0 \leq \lambda \leq \Lambda.$$

It is to be shown that $A \equiv K$. Differentiation produces

$$A_\lambda(t,y,\lambda) = k_{u\lambda}(t,y,\lambda,u(y,\lambda))$$

$$+ k_{uu}(t,y,\lambda,u(y,\lambda))\ u_\lambda(y,\lambda)$$

$$+ \int_0^1 k_u(t,y',\lambda,u(y',\lambda))\ K(y',y,\lambda)\ dy'$$

$$\tag{8.47}$$

$$+ \lambda \int_0^1 k_{u\lambda}(t,y',\lambda,u(y',\lambda))\ K(y',y,\lambda)\ dy'$$

$$+ \lambda \int_0^1 k_{uu}(t,y',\lambda,u(y',\lambda))\ u_\lambda(y',\lambda)K(y',y,\lambda)dy'$$

$$+ \lambda \int_0^1 k_u(t,y',\lambda,u(y',\lambda))\ K_\lambda(y',y,\lambda)\ dy'$$

and use of Eqs. (8.38) and (8.44) yields

$$A_\lambda(t,y,\lambda) = k_{u\lambda}(t,y,\lambda,u(y,\lambda))$$

$$+ k_{uu}(t,y,\lambda,u(y,\lambda))$$

$$\cdot \left\{ \Psi(y,\lambda) + \lambda \int_0^1 K(y,y',\lambda) \; \Psi(y',\lambda) \; dy' \right\}$$

$$+ \int_0^1 k_u(t,y',\lambda,u(y',\lambda)) \; K(y',y,\lambda) \; dy'$$

$$+ \lambda \int_0^1 k_{u\lambda}(t,y',\lambda,u(y',\lambda)) \; K(y',y,\lambda) dy'$$

$$+ \lambda \int_0^1 k_{uu}(t,y',\lambda,u,Y',\lambda)) \tag{8.48}$$

$$\cdot \left\{ \Psi(y',\lambda) + \lambda \int_0^1 K(y',y'',\lambda) \; \Psi(y'',\lambda) dy'' \right\}$$

$$\cdot K(y',y,\lambda) \; dy'$$

$$+ \lambda \int_0^1 k_u(t,y',\lambda,u(y',\lambda))$$

$$\cdot \left\{ Q(y',y,\lambda) + \lambda \int_0^1 K(y',y'',\lambda) Q(y'',y,\lambda) dy'' \right\} dy$$

It follows, through the realization that Q is the sum of th first five terms on the right hand side, that

$$A_\lambda(t,y,\lambda) = Q(t,y,\lambda)$$

$$+ \lambda \int_0^1 \left\{ k_u(t,y',\lambda,u(y',\lambda)) \right.$$

$$\tag{8.49}$$

$$+ \lambda \int_0^1 k_u(t,y'',\lambda,u(y'',\lambda)) \; K(y'',y',\lambda) \; dy'' \right\}$$

$$\cdot \; Q(y',y,\lambda) \; dy' \; .$$

Noticing that the quantity in square brackets is A, we find that the function A satisfies the differential equation

$$A_\lambda(t,y,\lambda) = Q(t,y,\lambda) + \lambda \int_0^1 A(t,y',\lambda) \; Q(y',y,\lambda) \; dy' \;,$$

$$0 \le t, y \le 1 \; . \quad (8.50)$$

The initial condition at $\lambda = 0$ is

$$A(t,y,0) = k_u(t,y,0,g(y,0)), \tag{8.51}$$

$$0 \le t, y \le 1 \; .$$

Comparing Equations (8.50) and (8.51) against Equations (8.44) and (8.45), and assuming uniqueness of solution, it is seen that

$$A(t,y,\lambda) = K(t,y,\lambda) \;, \tag{8.52}$$

or

$$K(t,y,\lambda) = k_u(t,y,\lambda,u(y,\lambda))$$

$$(8.53)$$

$$+ \lambda \int_0^1 k_u(t,y',\lambda,u(y',\lambda))\ K(y',y,\lambda)\ dy'\ ,$$

$$0 \le t,y \le 1,\quad 0 \le \lambda \le \Lambda\ ,$$

which is the integral equation satisfied by K .

Lastly, we introduce the function B to be

$$B(t,\lambda) = g(t,\lambda) + \lambda \int_0^1 k(t,y,\lambda,u(y,\lambda))\ dy\ , \qquad (8.54)$$

$$0 \le t \le 1,\quad 0 \le \lambda \le \Lambda.$$

We will show that $B \equiv u$.   Differentiation with respect to λ shows that

$$B_\lambda(t,\lambda) = g_\lambda(t,\lambda) + \lambda \int_0^1 k(t,y,\lambda,u(y,\lambda))\ dy$$

$$+ \lambda \int_0^1 k_\lambda(t,y,\lambda,u(y,\lambda))\ dy \qquad (8.55)$$

$$+ \lambda \int_0^1 k_u(t,y,\lambda,u(y,\lambda))\ u_\lambda(y,\lambda)\ dy\ .$$

Using Equations (8.39) and (8.38) we reduce Equation (8.55) to the form

$$B_\lambda(t,\lambda) = \Psi(t,\lambda)$$

$$+ \lambda \int_0^1 k_u(t,y,\lambda,u(y,\lambda)) \left\{ \Psi(y,\lambda) \right. \qquad (8.56)$$

$$\left. + \lambda \int_0^1 K(y,y',\lambda)\ \Psi(y',\lambda)\ dy' \right\} dy$$

or

$$B_\lambda(t,\lambda) = \Psi(t,\lambda)$$

$$+ \lambda \int_0^1 \Big\{ k_u(t,y,\lambda,u(y,\lambda))$$

(8.57)

$$+ \lambda \int_0^1 k_u(t,y',\lambda,u(y',\lambda)) \; K(y',y,\lambda) \; dy' \Big\}$$

$$\cdot \; \Psi(y,\lambda) \; dy \; .$$

Making use of Equation (8.53) we observe that

$$B_\lambda(t,\lambda) = \Psi(t,\lambda) + \lambda \int_0^1 K(t,y,\lambda) \; \Psi(y,\lambda) \; dy \; , \qquad (8.58)$$
$$0 \leq t \leq 1 \; .$$

Furthermore, from the definition of the function B in Equation (8.54) we have

$$B(t,0) = g(t,0), \qquad\qquad\qquad 0 \leq t \leq 1 \; . \qquad (8.59)$$

Keeping Equations (8.38) and (8.40) in mind we see that

$$B(t,\lambda) = u(t,\lambda) \; , \qquad\qquad 0 \leq t \leq 1, \quad 0 \leq \lambda \leq \Lambda; \qquad (8.60)$$

i.e.,

$$u(t,\lambda) = g(t,\lambda) + \lambda \int_0^1 k(t,y,\lambda,u(y,\lambda)) \; dy \; , \qquad (8.61)$$
$$0 \leq t \leq 1, \quad 0 \leq \lambda \leq \Lambda,$$

which completes the demonstration.

NUMERICAL ANALYSIS. The Cauchy system may be handled
numerically by using the method of lines. Much previous
experience indicates the efficacy of such an approach.
The general idea is to approximate integrals on the interval (0,1) by means of a quadrature formula of order N ,

$$\int_0^1 f(y) \; dy \; \cong \; \sum_{i=1}^{N} \; f(r_i) \; w_i \; . \tag{8.62}$$

In that way the differential integral equations are approximated by a system of ordinary differential equations
as closely as desired. By introducing

$$u(r_i,\lambda) = u_i(\lambda)$$
$$\Psi(r_i,\lambda) = \Psi_i(\lambda) \tag{8.63}$$
$$K(r_i,r_j,\lambda) = K_{ij}(\lambda) \; ,$$
$$i,j=1,2,\ldots,N, \quad 0\leq\lambda\leq\Lambda,$$

Equation (8.38) becomes, for example,

$$du_i(\lambda)/d\lambda = \Psi_i(\lambda) + \lambda \sum_{m=1}^{N} K_{im}(\lambda) \; \Psi_m(\lambda) \; w_m \; , \tag{8.64}$$

with similar equations approximating Equations (8.44), (8.39)
and (8.42) and the initial condition in Equations (8.40) and
(8.45). In all, there are $N^2 + N$ ordinary differential
equations subject to known initial conditions. This is
reasonable for $N \cong 50$.

DISCUSSION. We have shown the equivalence between a general
class of nonlinear integral equations and a Cauchy system.
Much remains to be studied analytically, especially with
regard to bifurcation phenomena.

8.4   LINEAR INTEGRAL EQUATIONS WITH NONLINEAR PARAMETERS

We consider Fredholm integral equations of the form

$$u(t) = g(t) + \lambda \int_0^1 k(t,y,\lambda) \, u(y) \, dy \,, \qquad (8.65)$$
$$0 \leq t \leq 1 \,.$$

Since the solution also depends upon the parameter λ, we
write

$$u = u(t,\lambda) \,, \qquad (8.66)$$

and equation (8.65) assumes the form

$$u(t,\lambda) = g(t) + \lambda \int_0^1 k(t,y,\lambda) \, u(y,\lambda) \, dy \,, \qquad (8.67)$$
$$0 \leq t \leq 1, \quad 0 \leq \lambda \leq \Lambda.$$

We shall now transform the integral equation (8.67) into an
initial value problem in which the parameter λ plays the
role of the time-like variable. Then we shall show the
computational utility of our considerations by applying the
method to the Lebedev and Ufliand integral equation.

DERIVATION. Through differentiation with respect to λ equation (8.67) becomes

$$u_\lambda(t,\lambda) = \int_0^1 k(t,y,\lambda)\ u(y,\lambda)\ dy$$

$$+ \lambda \int_0^1 k_\lambda(t,y,\lambda)\ u(y,\lambda)\ dy \qquad (8.68)$$

$$+ \lambda \int_0^1 k(t,y,\lambda)\ u_\lambda(y,\lambda)\ dy\ .$$

We view this as a Fredholm integral equation for the function u_λ . The first two terms on the right hand side of the above equation form the forcing term.

$$\Psi(t,\lambda) = \int_0^1 k(t,y,\lambda)\ u(y,\lambda)\ dy$$

$$\qquad (8.69)$$

$$+ \lambda \int_0^1 k_\lambda(t,y,\lambda)\ u(y,\lambda)\ dy\ ,$$

and equation (8.68) becomes

$$u_\lambda(t,\lambda) = \Psi(t,\lambda) + \lambda \int_0^1 k(t,y,\lambda)\ u_\lambda(y,\lambda)\ dy\ , \qquad (8.70)$$
$$0 \le t \le 1\ .$$

To solve equation (8.76) we introduce the resolvent kernel $K = K(t,y,\lambda)$, $0 \le t, y \le 1$, $0 \le \lambda \le \Lambda$. In terms of this function the solution of equation (8.70) is

$$u_\lambda(t,\lambda) = \Psi(t,\lambda) + \lambda \int_0^1 K(t,y,\lambda)\ \Psi(y,\lambda)\ dy\ , \qquad (8.71)$$

$$0 \le t \le 1, \quad 0 \le \lambda.$$

The resolvent kernel K itself satisfies the integral equation

$$K(t,y,\lambda) = k(t,y,\lambda) + \lambda \int_0^1 k(t,y',\lambda)\ K(y',y,\lambda)\ dy'$$

$$0 \le t, y \le 1, \quad 0 \le \lambda. \qquad (8.72)$$

To complete our Cauchy system we obtain a differential equation and initial condition for the resolvent kernel K. We differentiate equation (8.72) and find that

$$K_\lambda(t,y,\lambda) = k_\lambda(t,y,\lambda) + \int_0^1 k(t,y',\lambda)\ K(y',y,\lambda)\ dy'$$

$$+ \lambda \int_0^1 k_\lambda(t,y',\lambda)\ K(y',y,\lambda)\ dy' \qquad (8.73)$$

$$+ \lambda \int_0^1 k(t,y',\lambda)\ K(y',y,\lambda)\ dy'\ .$$

Equation (8.73) is viewed as an integral equation for the function K_λ having as its forcing term

$$Q(t,y,\lambda) = k_\lambda(t,y,\lambda) + \int_0^1 k(t,y',\lambda)\ K(y',y,\lambda)\ dy'$$

$$\qquad (8.74)$$

$$+ \lambda \int_0^1 k_\lambda(t,y',\lambda)\ K(y',y,\lambda)\ dy'\ ,$$

$$0 \le t, y \le 1, \quad 0 \le \lambda.$$

It follows that the solution of equation (8.73) is

$$K_\lambda(t,y,\lambda) = Q(t,y,\lambda) + \lambda \int_0^1 k(t,y',\lambda) \, Q(y',y,\lambda) \, dy',$$

$$0 \le t, y \le 1, \quad 0 \le \lambda, \quad (8.75)$$

which is the other desired differential equation.

The initial condition on the function u at $\lambda = 0$ is

$$u(t,0) = g(t) \, , \qquad\qquad\qquad 0 \le t \le 1 \, , \qquad (8.76)$$

which follows from equation (8.65). Similarly, the initial condition on the function R at $\lambda = 0$ is

$$K(t,y,0) = k(t,y,0) \, , \qquad\qquad 0 \le t, y \le 1 \qquad (8.77)$$

an immediate consequence of equation (8.72)

THE CAUCHY SYSTEM. It is now convenient to summarize the Cauchy system for the functions u and K. The initial conditions at $\lambda = 0$ are

$$u(t,0) = g(t) \, , \qquad\qquad\qquad 0 \le t \le 1 \, , \qquad (8.78)$$

and

$$K(t,y,0) = k(t,y,0) \, , \qquad\qquad 0 \le t, y \le 1 \, . \quad (8.79)$$

The differential equations for $\lambda \ge 0$ are

$$u_\lambda(t,\lambda) = \Psi(t,\lambda) + \lambda \int_0^1 K(t,y',\lambda) \; \Psi(y',\lambda) \; dy' \; , \tag{8.80}$$

and

$$K_\lambda(t,y,\lambda) = Q(t,y,\lambda) + \lambda \int_0^1 K(t,y',\lambda) \; Q(y',y,\lambda) \; dy' \; ,$$
$$0 \le t, y \le 1 \; , \tag{8.81}$$

where the auxiliary functions Ψ and Q are given by

$$\Psi(t,\lambda) = \int_0^1 \left\{ k(t,y',\lambda) + \lambda k_\lambda(t,y',\lambda) \right\} u(y',\lambda) \; dy' \tag{8.82}$$

and

$$Q(t,y,\lambda) = k_\lambda(t,y,\lambda)$$
$$+ \int_0^1 \left\{ k(t,y',\lambda) + \lambda k_\lambda(t,y',\lambda) \right\} K(y',y,\lambda) \; dy' \; ,$$
$$0 \le t, y \le 1, \quad 0 \le \lambda \le \Lambda. \tag{8.83}$$

NUMERICAL METHOD. To solve the initial value problem given above we employ the well-known method of lines. We approximate the integrals on the interval [0,1] by use of a quadrature formula

$$\int_0^1 f(y') \; dy' \cong \sum_{m=1}^M f(r_m) \; w_m \; . \tag{8.84}$$

Gaussian quadrature and Simpson's rule, for example, have this form. The result is that the differential-integral equations are transformed into a system of ordinary differential equations. We introduce the nomenclature

$$u(r_i, \lambda) = u_i(\lambda) \ ,$$

$$K(r_i, r_j, \lambda) = K_{ij}(\lambda) \ ,$$

$$\Psi(r_i, \lambda) = \Psi_i(\lambda) \ ,$$
(8.85)

$$Q(r_i, r_j, \lambda) = Q_{ij}(\lambda) \ ,$$

$$i, j = 1, 2, \ldots, M \ .$$

The approximating differential equations are then

$$u_i{}'(\lambda) = \Psi_i(\lambda) + \lambda \sum_{m=1}^{M} K_{im}(\lambda) \, \Psi_m(\lambda) \, w_m \ , \qquad (8.86)$$

and

$$K_{ij}{}'(\lambda) = Q_{ij}(\lambda) + \lambda \sum_{m=1}^{M} K_{im}(\lambda) \, Q_{mj}(\lambda) \, w_m \ , \qquad (8.87)$$

$$i, j = 1, 2, \ldots, M \ ,$$

where the primes stand for differentiation with respect to λ , and

$$\Psi_i(\lambda) = \sum_{m=1}^{M} \left\{ k(r_i, r_m, \lambda) + \lambda k_\lambda(r_i, r_m, \lambda) \right\} u_m(\lambda) \, w_m \ ,$$
(8.88)

$$Q_{ij}(\lambda) = k_\lambda(r_i, r_j, \lambda)$$

$$+ \sum_{m=1}^{M} \left\{ k(r_i, r_m, \lambda) + \lambda k_\lambda(r_i, r_m, \lambda) \right\} \qquad (8.89)$$

$$\cdot \; K_{mj}(\lambda) \; w_m \; .$$

The initial conditions at $\lambda = 0$ are

$$u_i(0) = g(r_i) \qquad\qquad (8.90)$$

and

$$K_{ij}(0) = k(r_i, r_j, 0) \; , \qquad\qquad (8.91)$$
$$i,j=1,2,\ldots,M \; .$$

Thus we have a system of $M^2 + M$ ordinary differential equations to integrate. This is a more or less routine task for $M \le 50$.

Should we desire the value of $u(t,\lambda)$ for $t \neq r_i$, $i=1,2,\ldots,M$, three methods immediately come to mind. In the first, we make use of the fact that $u(r_i,\lambda)$, $u(r_2,\lambda)$, $\ldots$, $u(r_m,\lambda)$ are known, and regard this as a standard problem in interpolation. If Gaussian quadrature has been employed, interpolation using orthogonal polynomials recommends itself. A second possibility consists in adjoining the differential equation

$$u_\lambda(t,\lambda) = \Psi(t,\lambda) + \lambda \sum_{m=1}^{M} K(t,r_m,\lambda) \; \Psi_m(\lambda) \; w_m \qquad (8.92)$$

and initial condition

$$u(t,0) = g(t) \tag{8.93}$$

to the basic system. Since values of $K(t,r_m,\lambda)$, $m=1,2,\ldots,M$, are required, we must also adjoin the differential equations and initial conditions

$$K_\lambda(t,r_j,\lambda) = Q(t,r_j,\lambda) \tag{8.94}$$

$$+ \lambda \sum_{m=1}^{M} K(t,r_m,\lambda)\, Q_{mj}(\lambda)\, w_m$$

$$K(t,r_j,0) = k(t,r_j,0)\, , \qquad\qquad j=1,2,\ldots,M. \tag{8.95}$$

We must also employ the formulas

$$\Psi(t,\lambda) = \sum_{m=1}^{M} \left\{ k(t,r_m,\lambda) + \lambda k_\lambda(t,r_m,\lambda) \right\} u_m(\lambda)\, w_m \tag{8.96}$$

and

$$Q(t,r_j,\lambda) = k_\lambda(t,r_j,\lambda)$$

$$+ \sum_{m=1}^{M} \left\{ k(t,r_m,\lambda) + \lambda k_\lambda(t,r_m,\lambda) \right\} \tag{8.97}$$

$$\cdot\, K_{mj}(\lambda)\, w_m\, ,$$

$$j=1,2,\ldots,M.$$

Thus, if we wish the value of u for N values of
$t \neq t_i$, i=1,2,...,M, we must adjoin N(M+1) additional
ordinary differential equations to the basic set of $M^2 + M$.
Use of this technique will be illustrated below.

Finally, we recall that u expressed in terms of
the resolvent is

$$u(t,\lambda) = g(t) + \lambda \int_0^1 K(t,y',\lambda) \ g(y') \ dy', \qquad (8.98)$$
$$0 \leq t \leq 1, \quad 0 \leq \lambda.$$

It follows that we need only integrate the ordinary differential equations for $K_{ij}(\lambda)$, i,j=1,2,...,M , if $t = r_k$ for
some k=1,2,...,M; or the ordinary differential equations
for $K_{ij}(\lambda)$, i,j=1,2,...,M, and $K(t,r_j,\lambda)$, j=1,2,...,M ,
if $t \neq r_k$, k=1,2,...,M. This technique will also be used
later.

8.5   APPLICATION TO LEBEDEV-UFLIAND EQUATION OF ELASTICITY

LEBEDEV-UFLIAND INTEGRAL EQUATION. In a famous paper
published in 1958, Lebedev and Ufliand considered the problem
of a circular punch penetrating into a finitely thick elastic
layer resting on a rigid foundation. They showed that the
required displacements and stresses could be expressed in
terms of a single auxiliary function of one independent
variable. This auxiliary function is the solution of a
Fredholm integral equation which, in the case of a flat punch,
assumes the form

$$\omega(\xi) = 1 + \frac{1}{\pi} \int_0^1 [K(\tau+\xi) + K(\tau-\xi)] \, \omega(\tau) \, d\tau, \qquad (8.100)$$
$$0 \leq \xi \leq 1 \; ,$$

where

$$K(r) = p \int_0^\infty \frac{\alpha + e^{-\alpha} \sinh \alpha}{\alpha + \sinh \alpha \cosh \alpha} \cos pr\alpha \, d\alpha \; , \qquad (8.101)$$

and the nondimensional parameter p is the ratio of the radius of the punch to the thickness of the layer.

It is desirable to be able to produce the function $\omega(\xi)$ $0 \leq \xi \leq 1$, for a range of values of the parameter p , say $0 < p \leq 4.0$. Some partial results are given in Table 2 of the Lebedev and Ufliand paper. They were produced by first computing values of the function $K = K(r)$, $0 \leq r \leq 2$, and then solving the integral equation (8.100) for various values of p . This involved approximating the integrals by use of a quadrature formula and solving the resulting systems of linear algebraic equations.

We also produce numerical solutions of the Lebedev-Ufliand equation. Our computational scheme is that described in the previous section. We obtain good agreement in short computing times.

Our general integral equation becomes identical to that of Lebedev and Ufliand when we set

$$g(t) = 1.0 \; , \hspace{5cm} 0 \leq t \leq 1 \; , \quad (8.102)$$

and

$$k(t,y,\lambda)$$

(8.103)

$$= (2/\pi) \int_0^\infty \frac{\alpha + e^{-\alpha} \sinh \alpha}{\alpha + \sinh \alpha \cosh \alpha} \cos \lambda t \alpha \cos \lambda y \alpha d\alpha$$

$$0 \leq t, y \leq 1, \quad 0 \leq \lambda.$$

To evaluate the kernel k we make the substitution

$$\beta = e^{-\alpha}, \qquad\qquad\qquad 0 \leq \alpha \leq \infty, \quad (8.104)$$

so that

$$k(t,y,\lambda) = (4/\pi) \int_0^1 \beta \frac{1-\beta^2 - 2\ln \beta}{1-\beta^4 - 4\beta^2 \ln\beta} \cos(\lambda y \ln\beta)$$

(8.105)

$$\cdot \cos(\lambda t \ln \beta) \, d\beta,$$

$$0 \leq t, y \leq 1, \quad 0 \leq \lambda.$$

The integral is then evaluated using Gaussian quadrature of order 13 on the interval [0,1].

In our first experiment we integrate equations (8.86) and (8.87) for M = 7, 9 and 11, and we adjoined equations (8.92) and (8.94) for t=0. 0, 0.1, 0.2,...,1.0. We use a fourth-order Adams-Moulton integration scheme with a Runge-Kutta start and a step-size of 0.01. The agreement is at least two significant places, and is usually four figures. These results demonstrate that M = 7 is adequate, though undoubtedly a smaller value would suffice. In our further experiments we use M = 7.

In the second experiment, we set M = 7 and use integration step sizes of H = .01 and .005. The results show that an integration step-size of .10 is adequate for agreement to one part in 3600, though undoubtedly a larger one would suffice.

In our third experiment we calculate $u(r_i, \lambda)$, i=1,2,..., M = 7 , using two different methods. The first, called the I.V. method in Table 8.1, uses the full set of differential equations described in Section 8.4 The second uses the calculation of the resolvent and formula (8.98). The agreement is excellent, and the results are shown in Table 8.1.

Table 8.1

The Function $u(t,\lambda)$ Using Cauchy System and Resolvent Kernels

(Shown at Gaussian Quadrature Abscissas)

Abscissa	$\lambda = 2.0$		$\lambda = 4.0$	
	I.V. Method	Resolvent Method	I.V. Method	Resolvent Method
.025	4.329	4.329	8.341	8.340
.129	4.300	4.300	8.271	8.270
.297	4.168	4.168	7.977	7.977
.500	3.864	3.864	7.320	7.320
.703	3.416	3.416	6.133	6.132
.871	2.973	2.973	4.746	4.745
.975	2.687	2.687	3.838	3.838

Table 8.2 gives values of the function u for
t = 0.0, 0.1,..., 1.0 and λ = 1.0, 2.0, 3.0 and 4.0.
Comparison against Table 8.2 of Lebedev and Ufliand is
excellent for $\lambda \leq 2.0$, the discrepancy being no greater than
one-quarter percent. For $\lambda > 2.0$, our values are new. We
use M = 7, N = 11, and H = .01 to produce our tables.
Thus we are integrating simultaneously 7^2 + 7 + (7+1)11 =
144 ordinary differential equations with known initial
conditions. Execution time is several minutes.

Table 8.2

Values of the Function $u(t,\lambda)$

t	λ = 1.0	λ = 2.0	λ = 3.0	λ = 4.0
0.0	2.356	4.331	6.342	8.344
0.1	2.351	4.312	6.313	8.300
0.2	2.335	4.257	6.226	8.172
0.3	2.310	4.164	6.080	7.970
0.4	2.274	4.033	5.868	7.694
0.5	2.230	3.864	5.580	7.320
0.6	2.178	3.660	5.213	6.813
0.7	2.120	3.423	4.772	6.154
0.8	2.056	3.165	4.269	5.360
0.9	1.987	2.892	3.724	4.487
1.0	1.916	2.617	3.172	3.625

In Table 8.3 we show values of the resolvent kernel $K = K(t,y,\lambda)$. They are given at the seven Gaussian abscissas for t and y and at $\lambda = 1.0, 2.0, 3.0$ and 4.0. The table displays the expected symmetry of K in the arguments t and y . This symmetry could have been used to roughly cut the number of ordinary differential equations in half. The rows in each of the four matrices of Table 8.3 correspond to the values of t :

$$t = .025, .129, .297, .500, .703, .871, .975 ;$$

Similarly for the columns which correspond to the values of y .

Table 8.3

Resolvent Kernel at Gaussian Abscissas

$\lambda = 1.0$

1.591	1.577	1.519	1.396	1.233	1.083	.988
1.577	1.564	1.507	1.387	1.227	1.079	.986
1.519	1.507	1.456	1.349	1.203	1.064	.976
1.396	1.387	1.349	1.265	1.146	1.029	.952
1.233	1.227	1.203	1.146	1.061	.972	.910
1.083	1.079	1.064	1.029	.972	.908	.862
.988	.986	.976	.952	.910	.862	.825

$\lambda = 2.0$

2.348	2.295	2.085	1.706	1.295	.983	.812
2.295	2.249	2.057	1.699	1.299	.988	.817
2.085	2.057	1.933	1.661	1.309	1.009	.835
1.706	1.699	1.661	1.534	1.290	1.029	.863
1.295	1.299	1.309	1.290	1.180	1.008	.877
.983	.988	1.009	1.029	1.008	.928	.844
.812	.817	.835	.863	.877	.844	.792

$\lambda = 3.0$

2.830	2.721	2.329	1.758	1.244	.878	.678
2.721	2.637	2.308	1.773	1.257	.885	.688
2.329	2.308	2.179	1.811	1.313	.927	.719
1.758	1.773	1.811	1.732	1.398	1.006	.768
1.244	1.257	1.313	1.398	1.318	1.052	.844
.878	.885	.927	1.006	1.052	.984	.859
.678	.688	.719	.768	.844	.859	.795

$\lambda = 4.0$

3.175	2.991	2.418	1.751	1.206	.817	.608
2.991	2.874	2.424	1.780	1.225	.820	.599
2.418	2.424	2.339	1.887	1.300	.850	.626
1.751	1.780	1.887	1.901	1.458	.975	.701
1.206	1.225	1.300	1.458	1.464	1.081	.776
.817	.820	.850	.975	1.081	1.008	.855
.608	.599	.626	.701	.776	.855	.818

8.6   PARAMETER IMBEDDING FOR THE AMBARZUMIAN INTEGRAL
       EQUATION

INTRODUCTION. In two papers published in 1942 and 1943,
Ambarzumian introduced the nonlinear integral equation

$$\phi(\eta) = 1 + \frac{\lambda}{2} \eta \phi(\eta) \int_0^1 \frac{\phi(\xi)}{\eta+\xi} \, d\xi \, , \qquad (8.106)$$

$$0 \le \eta \le 1, \quad 0 \le \lambda \le 1 \, ,$$

into the study of radiative transfer. Since then it has
been the object of much study, for it is the key to multiple
scattering in semi-infinite atmospheres. In this section,
we reduce Eq. (8.106) to a Cauchy system in which the parameter λ is the independent variable. We describe the
results of some numerical experiments based on our initial
value method.

By dividing both sides of Eq. (8.106) by $\phi(\eta)$, the
equation assumes the form

$$1 - [\phi(\eta)]^{-1} = \lambda \int_0^1 \frac{\eta}{2(\eta+\xi)} \, \phi(\xi) \, d\xi \, , \qquad (8.107)$$

$$0 \le \eta \le 1 \, .$$

To emphasize that the function ϕ depends upon λ as well
as η we shall write

$$\phi = \phi(\eta,\lambda) \, , \qquad\qquad 0 \le \eta, \lambda \le 1 \, , \quad (8.108)$$

and Eq. (8.107) becomes

$$1 - [\phi(\eta,\lambda)]^{-1} = \lambda \int_0^1 \frac{\eta}{2(\eta+\xi)} \phi(\xi,\lambda) \, d\xi \, , \qquad (8.109)$$

For simplicity, and more generality, we shall write the above equation as

$$F(u(t,\lambda),t) = \lambda \int_0^1 k(t,y) \, u(y,\lambda) \, dy \, , \qquad (8.110)$$
$$0 \le t \le 1 \, ,$$

where F is, in general, a nonlinear function of the argument u. Should F be linear in u,

$$F(u(t,\lambda),t) = u(t,\lambda) - g(t) \, , \qquad (8.111)$$

then Eq. (8.110) reduces to the usual linear Fredholm integral equation equation of the second kind. We assume that Eq. (8.110) possesses a real solution for each value of λ in the interval $(0,\Lambda)$ and that it is differentiable in λ. We now derive the Cauchy system.

DERIVATION OF CAUCHY SYSTEM. Differentiate both sides of Eq. (8.110) with respect to λ to obtain the relation

$$F_u(u(t,\lambda),t) \, u_\lambda(t,\lambda) = \int_0^1 k(t,y) \, u(y,\lambda) \, dy$$

$$(8.112)$$

$$+ \lambda \int_0^1 k(t,y) \, u_\lambda(y,\lambda) \, dy \, .$$

This is regarded as a linear Fredholm integral equation of
the second kind for the function u_λ . It has the form

$$w(t,\lambda) = \frac{g(t,\lambda)}{F_u(t(t,\lambda),t)} + \lambda \int_0^1 \frac{k(t,y)}{F_u(u(t,\lambda),t)} w(y,\lambda)\, dy \; ,$$

$$0 \leq t \leq 1 \; , \qquad (8.113)$$

the solution of which may be written as

$$w(t,\lambda) = \frac{g(t,\lambda)}{F_u(t(t,\lambda),t)} + \lambda \int_0^1 K(t,y,\lambda)\, \frac{g(y,\lambda)}{F_u(u(y,\lambda),y)}\, dy \; ,$$

$$(8.114)$$

where we have introduced the resolvent kernel K . As is
known, this function satisfies the Fredholm integral equation

$$K(t,y,\lambda) = \frac{k(t,y)}{F_u(u(t,\lambda),t)} + \lambda \int_0^1 \frac{k(t,y')}{F_u(u(t,\lambda),t)} K(y',y,\lambda)\, dy'$$

$$0 \leq t,y,\lambda \leq 1 \; . \qquad (8.115)$$

Next we write

$$\Psi(t,\lambda) = \int_0^1 k(t,y)u(y,\lambda)\, dy \; , \qquad\qquad (8.116)$$

$$0 \leq t,\lambda \leq 1 \; .$$

Then the solution of Eq. (8.112)

$$u_\lambda(t,\lambda) = \frac{\Psi(t,\lambda)}{F_u(u(t,\lambda),t)} + \lambda \int_0^1 K(t,y,\lambda)\, \frac{\Psi(y,\lambda)}{F_u(u(y,\lambda),t)}\, dy,$$

$$(8.117)$$

which is one of the basic equations of our Cauchy system.
Now we must obtain a differential equation for the resolvent
kernel K .   Equation (8.115) is rewritten as

$$F_u(u(t,\lambda),t) \; K(t,y,\lambda) = k(t,y) + \lambda \int_0^1 k(t,y') \; K(y',y,\lambda) dy',$$

$$(8.118)$$

and, through differentiation becomes

$$F_{uu}(u(t,\lambda),t) \; u_\lambda(t,\lambda) K(t,y,\lambda) + F_u(u(t,\lambda),t) K_\lambda(t,y,\lambda)$$

$$= \int_0^1 k(t,y') \; K(y',y,\lambda) \; dy' \qquad\qquad (8.119)$$

$$+ \lambda \int_0^1 k(t,y') \; K_\lambda(y',y,\lambda) \; dy' \; .$$

Let us introduce the new function Q by means of the
definition

$$Q(t,y,\lambda) = \int_0^1 k(t,y') \; K(y',y,\lambda) \; dy'$$

$$- F_{uu}(u(t,\lambda),t) \left\{ \frac{\Psi(t,\lambda)}{F_u(u(t,\lambda),t)} \right.$$

$$+ \lambda \int_0^1 K(t,y',\lambda) \; \frac{\Psi(y',\lambda)}{F_u(u(y',\lambda),y')} \; dy' \left. \right\} \qquad (8.120)$$

$$\cdot \; K(t,y,\lambda) \; .$$

The solution of equation (8.119), viewed as a linear Fredholm integral equation for the function K_λ , is

$$K_\lambda(t,y,\lambda) = \frac{Q(t,y,\lambda)}{F_u(u(t,\lambda),t)} \tag{8.121}$$

$$+ \lambda \int_0^1 K(t,y',\lambda) \frac{Q(y',y,\lambda)}{F_u(u(y',\lambda),y')} \, dy' \; .$$

This is the other basic differential equation.

The initial conditions at $\lambda = 0$ on the functions u and K are easily obtained. From Eq. (8.110) we see that

$$F(u(t,0),t) = 0 \; , \qquad\qquad 0 \le t \le 1 \; , \tag{8.122}$$

and we denote the relevant root of this equation by $c = c(t)$,

$$u(t,0) = c(t) \; , \qquad\qquad 0 \le t \le 1 \; . \tag{8.123}$$

From Eq. (8.115) we see that

$$K(t,y,0) = \frac{k(t,y)}{F_u(c(t),t)} \; , \qquad\qquad 0 \le t, y \le 1 \; . \tag{8.124}$$

The needed equations have now been obtained, and we may state the Cauchy system satisfied by the function u and the auxiliary function K .

STATEMENT OF CAUCHY SYSTEM. Let the nonlinear integral
equation (eq. (8.110) possess a solution for each value of
λ in the interval $(0,\Lambda)$. Also let the solution be differentiable in λ . Then the function u and an auxiliary
function K satisfy the differential equations

$$u_\lambda(t,y) = \frac{\Psi(t,\lambda)}{F_u(u(t,\lambda),t)}$$

$$+ \lambda \int_0^1 K(t,y',\lambda) \frac{\Psi(y',\lambda)}{F_u(u(y',\lambda),y')} dy' \quad , \tag{8.125}$$

$$K_\lambda(t,y,\lambda) = \frac{Q(t,y,\lambda)}{F_u(u(t,\lambda),t)}$$

$$+ \lambda \int_0^1 K(t,y',\lambda) \frac{Q(y',y,\lambda)}{F_u(u(y',\lambda),y')} dy' \quad , \tag{8.126}$$

where

$$\Psi(t,\lambda) = \int_0^1 k(t,y') u(y',\lambda) dy' \tag{8.127}$$

and

$$Q(t,y,\lambda) = \int_0^1 k(t,y') K(y',y,\lambda) dy' - F_{uu}(u(t,\lambda),t)$$

$$\cdot \left\{ \frac{\Psi(t,\lambda)}{F_u(u(t,\lambda),t)} + \right. \tag{8.128}$$

$$+ \lambda \int_0^1 K(t,y',\lambda) \frac{\Psi(y',\lambda)}{F_u(u(y',\lambda),y')} dy' \left. \right\} K(t,y,\lambda) \ .$$

The initial conditions at $\lambda = 0$ are

$$u(t,0) = c(t), \qquad\qquad\qquad 0 \leq t \leq 1 , \qquad (8.129)$$

and

$$K(t,y,0) = \frac{k(t,y)}{F_u(c(t),t)} , \qquad\qquad 0 \leq t, \quad y \leq 1 , \qquad (8.130)$$

where c is a root of the nonlinear equation

$$F(c(t),t) = 0 . \qquad\qquad\qquad\qquad (8.131)$$

NUMERICAL CONSIDERATIONS. Numerical solution of the Cauchy system described is obtained through use of the well-known method of lines. Integrals on the interval $(0,1)$ are approximated by a finite sum

$$\int_0^1 f(y) \, dy \cong \sum_{m=1}^{N} f(r_m) \, w_m , \qquad\qquad (8.132)$$

and the system of differential-integral equations is approximated by a system of ordinary differential equations. We write

$$u(r_i,\lambda) = u_i(\lambda) , \qquad\qquad\qquad (8.133)$$

$$K(r_i,r_j,\lambda) = K_{ij}(\lambda) \qquad\qquad\qquad (8.134)$$

$$\Psi(r_i, \lambda) = \Psi_i(\lambda) \tag{8.135}$$

$$Q(r_i, r_j, \lambda) = Q_{ij}(\lambda) \ , \tag{8.136}$$

$$i,j=1,2,\ldots,N, \quad 0\leq\lambda\leq\Lambda.$$

Equations (8.125) - (8.128) are approximated by the system

$$\dot{u}_i(\lambda) = \frac{\Psi_i(\lambda)}{F_u(u_i(\lambda), r_i)}$$

$$+ \lambda \sum_{n=1}^{N} K_{in}(\lambda) \frac{\Psi_n(\lambda)}{F_u(u_n(\lambda), r_n)} w_n \ , \tag{8.137}$$

$$\dot{K}_{ij}(\lambda) = \frac{Q_{ij}(\lambda)}{F_u(u_i(\lambda), r_i)}$$

$$+ \lambda \sum_{n=1}^{N} K_{in}(\lambda) \frac{Q_{nj}(\lambda)}{F_u(u_n(\lambda), r_n)} w_n \ , \tag{8.138}$$

$$\Psi_i(\lambda) = \sum_{n=1}^{N} k(r_i, r_n) u_n(\lambda) w_n \ , \tag{8.139}$$

$$Q_{ij}(\lambda) = \sum_{n=1}^{N} k(r_i, r_n) K_{nj}(\lambda) w_n - F_{uu}(u_i(\lambda), r_i) K_{ij}(\lambda)$$

$$\cdot \left[\frac{\Psi_i(\lambda)}{F_u(u_i(\lambda), r_i)} + \lambda \sum_{n=1}^{N} K_{in}(\lambda) \frac{\Psi_n(\lambda)}{F_u(u_n(\lambda), r_n)} w_n \right] \ ,$$

$$i,j=1,2,\ldots,N \ . \tag{8.140}$$

and the dots in Eqs. (8.137) and (8.138) represent differentiation with respect to the independent variable λ. The
initial conditions at $\lambda = 0$ are

$$u_i(0) = c(r_i) , \tag{8.141}$$

and

$$K_{ij}(0) = \frac{k(r_i,r_j)}{F_u(c(r_i),r_i)} , \tag{8.142}$$

$$i,j=1,2,\ldots,N,$$

where $c(r_i)$ is a root of the equation

$$\beta_i = c(r_i) , \tag{8.143}$$

$$F(\beta_i,r_i) = 0 , \qquad\qquad i=1,2,\ldots,N . \tag{8.144}$$

Thus $N + N^2$ ordinary differential equations subject to
initial conditions have to be integrated numerically. This
is rather routine for systems of order 10^3 or 10^4 .

It is frequently desirable to be able to determine
the value of $u(t,\lambda)$ at a value of $t \neq r_i$, $i=1,2,\ldots,N$
(in particular, t need not lie in the interval $(0,1)$.)
This is done by introducing the new variables $u(t,\lambda)$,
$\Psi(t,\lambda)$, $K(t,r_n,\lambda)$ and $Q(t,r_n,\lambda)$, $n=1,2,\ldots,N$, which we
shall denote by $u(\lambda)$, $\Psi(\lambda)$, $K_n(\lambda)$ and $Q_n(\lambda)$ respectively, $n=1,2,\ldots,N$. The differential equations and finite
relations to be adjoined to the equations above are

$$\dot{u}(\lambda) = \frac{\Psi(\lambda)}{F_u(u(\lambda),t)}$$

$$\qquad\qquad (8.145)$$

$$+ \lambda \sum_{n=1}^{N} K_n(\lambda) \frac{\Psi_n(\lambda)}{F_u(u_n(\lambda),r_n)} w_n \;,$$

$$\dot{K}_j(\lambda) = \frac{Q_j(\lambda)}{F_u(u(\lambda),t)}$$

$$\qquad\qquad (8.146)$$

$$+ \lambda \sum_{n=1}^{N} K_n(\lambda) \frac{Q_{nj}(\lambda)}{F_u(u_n(\lambda),r_n)} w_n \;,$$

$$\Psi(\lambda) = \sum_{n=1}^{N} k(t,r_n)\, u_n(\lambda)\, w_n \;, \qquad\qquad (8.147)$$

$$Q_j(\lambda) = \sum_{n=1}^{N} k(t,r_n) K_{nj}(\lambda)\, w_n$$

$$- F_{uu}(u(\lambda),t)\, K_j(\lambda) \qquad\qquad (8.148)$$

$$\cdot \left[\frac{\Psi(\lambda)}{F_u(u(\lambda),t)} + \lambda \sum_{n=1}^{N} K_n(\lambda) \frac{\Psi_n(\lambda)}{F(u_n(\lambda),r_n)} w_n \right],$$

$$j=1,2,\dots,N \;.$$

The initial conditions at $\lambda = 0$ are

$$u(0) = \alpha \qquad\qquad (8.149)$$

where α is a root of

$$F(\alpha,t) = 0 \qquad\qquad (8.150)$$

and

$$K_n(0) = \frac{k(t,r_n)}{F_u(\alpha,t)} , \qquad\qquad n=1,2,\ldots,N . \qquad (8.151)$$

Thus each additional point t for which values of $u(t,\lambda)$
are desired leads to $N + 1$ additional ordinary differential
equations to be added to the basic set of $N^2 + N$. If
there are K additional values of t , then the total number
of ordinary differential equations is $N^2 + N + (N + 1) K$.

RESULTS OF NUMERICAL EXPERIMENTS. To test the efficacy of
the proposed method, we solve the Ambarzumian integral equation by its use. In this instance, we have

$$F(u(t,\lambda),t) = 1 - [u(t,\lambda)]^{-1} , \qquad\qquad (8.152)$$

$$F_u(u(t,\lambda),t) = [u(t,\lambda)]^{-2} , \qquad\qquad (8.153)$$

$$F_{uu}(u(t,\lambda),t) = -2[u(t,\lambda)]^{-3} , \qquad\qquad (8.154)$$

$$k(t,y) = (t/2)/(t+y) , \qquad\qquad (8.155)$$

$$d(t) = 1 , \qquad\qquad 0 \le t,y,\lambda \le 1 . \qquad (8.156)$$

In our first numerical experiment, we employ Gaussian
quadrature of order $N = 7$ and use a fourth-order Adams-
Moulton integration scheme with an integration step size of
0.01. Much previous experience indicated that this would be

appropriate. We produce the function $\phi(\eta,\lambda)$ for the seven values of η appropriate for Gaussian quadrature on the interval $(0,1)$ and at $\eta = 0.0, 0.1, 0.2,\ldots,1.0$. In all we are integrating $7^2 + 7 + 11 (7 + 1) = 144$ simultaneous ordinary differential equations with known initial conditions. For comparison purposes, we use the table given by Ivanov on page 160 and by Abhyankar and Fymat.

In view of the bifurcation point at $\lambda = 1.0$, we wonder what the computational results would be like as λ approaches this value.

The calculations are done on an IBM System 360-44. For $\lambda \leq 0.95$ we are in perfect agreement with Ivanov. As λ tends to 1.0, the numerical solution remains bounded, and at $\lambda = 1.0$ the greatest discrepancy between our results (at $\mu = 1.0$) amounts to five percent. By using an alternative form of the Ambarzumian equation, we are confident this discrepancy can be eliminated.

This experiment, with execution time of less than one minute, shows the complete feasibility of the proposed method. It must be borne in mind that the calculations produce values of $\phi(\eta,\lambda)$ for eighteen values of η and 100 values of λ with no use of successive approximations or series expansions.

Next, we do a run in which we ask for the integration to proceed beyond $\lambda = 1.01$ with an overflow. We also calculate values of ϕ for $\eta = 1.1, 1.2,\ldots,1.5$.

In the third numerical experiment, we put $N = 9$ and use an integration step size of 0.005 to improve the accuracy. We find

$$\phi = (1.0, 0.995) = 2.5835 ,$$

whereas Ivanov gives

$$\phi(1.0,0.995) = 2.5873 \ .$$

In our final experiment, to cut the running time drastically, we put $N = 3$ and use an integration step size in λ of 0.01. At $\lambda = 0.95$ we are still in perfect agreement with Ivanov, while at $\lambda = 1.0$ there is a maximum discrepancy at $\eta = 1.0$ of 5 percent.

DISCUSSION. In this section we have presented a new method for the analytical and computational study of the nonlinear integral equation (8.110), and we have shown its numerical efficiency by solving the Ambarzumian integral equation. The main advantage of the method is that it automatically does a parameter study in the parameter λ . Much remains to be done in the investigation of bifurcation phenomena, in improving the computational efficiency, and in applying the method to other nonlinear integral equations.

8.7   EIGENVALUE PROBLEMS

We wish to solve the homogeneous equation

$$U(t) = \lambda \int_0^1 k(t,y) \, U(y) \, dy . \tag{8.157}$$

Fredholm showed that the resolvent $K(t,y;\lambda)$ of the kernel $k(t,y)$ could be expressed as

$$K(t,y;\lambda) = \frac{D(t,y;\lambda)}{D(\lambda)} , \tag{8.158}$$

where $D(t,y;\lambda)$ and $D(\lambda)$ are entire functions of λ. He also showed that $D(t,y;\lambda)$ and $D(\lambda)$ are related by the differential equation

$$D_\lambda(\lambda) = -\int_0^1 D(y',y';\lambda) \, dy' \tag{8.159}$$

and

$$D(0) = 1 , \tag{8.160}$$

where the subscript denotes differentiation with respect to λ. The eigenvalues are the roots of the equation

$$D(\lambda) = 0 . \tag{8.161}$$

At a simple root $\lambda = \lambda^*$ an eigenfunction is given by $D(t,y;\lambda^*)$ for any y.

Earlier it has been shown that the resolvent kernel $K(t,y;\lambda)$ satisfies the Cauchy system

$$K_\lambda(t,y;\lambda) = \int_0^1 K(t,y';\lambda) \; K(y',y;\lambda) \; dy' \; , \qquad (8.162)$$

subject to the initial condition

$$K(t,y;0) = k(t,y) \; , \qquad\qquad\qquad\qquad (8.163)$$
$$0 \leq t,y \leq 1.$$

If desired this system can be easily solved numerically to actually produce the resolvent K.

It is then a simple matter to show that the function $D(t,y;\lambda)$ satisfies the differential equation

$$D_\lambda(t,y;\lambda) = [D(\lambda)]^{-1}\left\{ \int_0^1 D(t,y;\lambda) \; D(y',y;\lambda) \; dy' \right.$$

$$\qquad\qquad\qquad\qquad\qquad\qquad (8.164)$$

$$\left. - D(t,y;\lambda) \int_0^1 D(y',y',\lambda) \; dy' \right\}$$

and we find that the initial condition is

$$D(t,y;0) = k(t,y) \; . \qquad\qquad\qquad (8.165)$$

From the theory of a complex variable, it is known that the number of zeros, N, of $D(\lambda)$ (each counted with proper multiplicity) in a closed contour C, with none on C, is given by

$$N = \frac{1}{2\pi i} \oint_C \frac{D_\lambda(\lambda)}{D(\lambda)} \, d\lambda \; . \tag{8.166}$$

We shall propose a numerical scheme for evaluating N.

Assume that there are no roots on the unit circle. To find the number of eigenvalues enclosed by the unit circle in the λ-plane, we numerically integrate the initial value problem for the functions $D(t,y;\lambda)$ and $D(\lambda)$, to $\lambda = P$, a point on the unit circle. In particular, this gives the values of $D(P)$ and $D_\lambda(P)$. Then we adjoin the differential equation

$$\frac{d\Psi}{d\lambda} = \frac{1}{2\pi i} \frac{D_\lambda(\lambda)}{D(\lambda)} = \frac{1}{2\pi i} \frac{-\int_0^1 D(y',y';\lambda) \, dy'}{D(\lambda)} \; , \tag{8.167}$$

and the initial condition

$$\Psi(P) = 0 \; , \tag{8.168}$$

to the Cauchy system for $D(t,y;\lambda)$ and $D(\lambda)$ described above. We now integrate this new system around the unit circle. It follows that the increment of Ψ, which must be zero or a positive integer, is the number of roots enclosed by the unit circle.

In an obvious manner, we can further localize the position of a root to various sectors, as when finding the roots of a polynomial numerically. The formula

$$\frac{1}{2\pi i} \oint_C \lambda^k \frac{D_\lambda(\lambda)}{D(\lambda)} \, d\lambda = \sum_{i=1}^{M} \lambda_i^{\ k} \ , \tag{8.169}$$

$$k=1,2,\ldots,$$

where $\lambda_1, \lambda_2,\ldots,\lambda_M$ are the roots enclosed by the unit circle, can be used to aid in the determination of both loca-tions and multiplicities. Many obvious possibilities exist for finding the roots outside the unit circle.

We shall present two numerical examples to illustrate the efficacy of the method. A Gaussian quadrature scheme of order N is used to approximate the intergrals, and a stand-ard fourth order Runge-Kutta scheme employing complex arith-metic is used to integrate numerically the resulitng system of differential equations. For a quadrature scheme of order N, we get N^2 differential equations for $D(t,y,\lambda)$ and one equation for each of D and Ψ. Additional equations may be added by letting

$$\frac{d}{d\lambda} \Omega_k = \frac{1}{2\pi i} \lambda^k \frac{D_\lambda(\lambda)}{D(\lambda)} \ , \qquad\qquad k=1,2,\ldots. \tag{8.170}$$

As our first example we shall consider the integral equation

$$U(t) = \lambda \int_0^1 x \, t \, U(x) \, dx \ , \tag{8.171}$$

$$0 \leq t \leq 1.$$

Since this kernel is degenerate and symmetric it is easy to show that it has one real eigenvalue at $\lambda = 3$. For the numerical experiments we let N = 5 and chose to integrate from the origin along the positive imaginary axis, using a step size of Δr = 0.05, to the circle of unit radius. At this point we adjoin two additional differential equations for Ψ and Ω. We then integrate, using a step size of $\Delta\theta = \frac{1^\circ}{2}$, counterclockwise around the unit circle. Upon returning to the point (0,1) we obtain the following results (using the notation 7.5666-8 = 7.5666 x 10^{-8}) ,

$$\Psi = (7.5666\text{-}8, - 2.2374\text{-}7) ,$$

$$\Omega_1 = (2.2637\text{-}7, - 7.6561\text{-}8) ,$$

indicating that the unit circle contains no zeros. The process is then repeated for a circle of radius four. The results are

$$\Psi = (1.0000, - 3.5642\text{-}7) ,$$

$$\Omega_1 = (2.9824, - 1.3037\text{-}2) ,$$

These results indicate that the circle of radius four encloses one eigenvalue, and its position is approximately at λ = (3,0). These results could easily be improved upon by increasing the number of quadrature points and decreasing the integration step size.

The second experiment is for the equation

$$U(t) = \lambda \int_0^1 i \; x \; t \; U(x) \; dx . \qquad\qquad (8.172).$$

This kernel has only one eigenvalue, and it is located at (0,-3) in the λ-plane. The above integration procedure is repeated in this case, and we obtain:

$$\text{radius one} \quad \Psi = (1.8672\text{-}7, 8.1492\text{-}10)$$
$$\Omega_1 = (\text{-}8.2814\text{-}10, 1.8892\text{-}7);$$

$$\text{radius four} \quad \Psi = (1.0000, 1.8623\text{-}9),$$
$$\Omega_1 = (\text{-}3.0542\text{-}2, 2.9998).$$

We have described a technique for calculating the comples eigenvalues for integral equations with general kernels. The method involves the combination of classical results of integral equations and complex variable theory with an imbedding method for converting Fredholm integral equations into initial valued differential equation systems. The technique is a flexible and accurate method for obtaining the complex eigenvalues of general kernels.

8.8   SUMMARY

We have given a complete description of parameter imbedding for integral equations. We have separately discussed linear integral equations with parameters appearing linearly, equations nonlinear in both the parameters and the solutions, and equations linear in the solutions but nonlinear in the parameters.

We have discussed numerical considerations and have presented selected results in some applications. We have also described a method for finding eigenvalues.

The reader will probably have recognized that this chapter forms the genesis of an earlier chapter in which we

treated the case of a general kernel by a transformation to
the interval [0,1].

EXERCISES

1. In Sec. 8.4, we derived imbedding equations (8.11) and
    (8.15) for $u(t,\lambda)$ and $K(t,y,\lambda)$ which are solutions
    of integral equations (8.1) and (8.7) respectively. In
    an alternate derivation based on the superposition prin-
    ciple, derive Eqs. (8.11) and (8.15) by noting that the
    forcing function in Eq. (8.7) for the resolvent is $k(t,y)$.
2. Give the validation for the Cauchy system for Eq. (8.65).
3. Show that this is a special case of imbedding for non-
    linear parameters.
4. Derive and validate parameter imbedding equations for
    systems of integral equations.

REFERENCES

1. N. Lebedev and Ia. Ufliand, "Osesimmetrichnaia Kontaknaia
    Zadacha dlia Uprugovo Sloia," *Prik. Mat. i Mekh,* v. 22,
    no. 3 (1958), pp. 320-326.
2. J. Buell, A. Fymat and R. Kalaba, "An Initial Value Method
    for the Ambarzumian Integral Equation," *J. Quant. Spectros.
    Rad. Trans.* v. 12, no. 5 (1972), pp. 769-776.
3. V. Ivanov, *Perenos Izlucheniya: Spektry Nebesnykh Tel.*
    Moskva (1969), p. 160.
4. H. Kagiwada, R. Kalaba and C. C. Yang, "Reduction of a
    Class of Nonlinear Integral Equations to a Cauchy System,"
    J. Math. Phys. v. 13, no. 2 (1972), pp. 228-231.

5. R. Kalaba and E. Zagustin, "Reduction of Fredholm Integral
    Equations to Cauchy Systems," *J. Franklin Institute*,
    v. 293, no. 4 (1972), pp. 277-283.
6. R. Kalaba and M. Scott, "An Initial Value Method for
    Integral Operators - IV. Complex-Valued Kernels of Laser
    Theory," *J. Quant. Spectrosc. Rad. Trans.*, v. 13 (1973),
    pp. 509-515.
7. R. Kalaba, A. Yakush and E. Zagustin, "A New Method for
    the Lebedev-Ufliand Integral Equation for Contact Problems
    of Elasticity," *J. of Elasticity*, v.2 (1972), pp. 181-189.

CHAPTER 9

RADIATIVE TRANSFER

9.1   INTRODUCTION

A basic problem in radiative transfer is the determination of the internal field due to multiple scattering in a medium illuminated by monodirectional radiation. The present book came about as a result of the authors' interest in problems of multiple scattering of radiation. Our background in the early 1960s made us firm believers in the electronic computer's ability to solve initial value problems. We were also aware that certain differential equations could be derived when a scattering medium is increased in thickness by a small amount. We then succeeded in obtaining the complete initial value problem for the basic integral equation of radiative transfer using an entirely physical argument. Further analytical work led to our theory for displacement kernels and to the other types of kernels and integral equations which we have discussed. In this chapter we show how we first derived the Cauchy system for the radiative transfer problem. The reader may wish to re-derive the equations

303

analytically. Selected numerical results for various interesting functions are presented in graphical and tabular form.

The second part of this chapter reveals the importance
of the b and h functions. The b and h functions are physically the internal intensities due to isotropic illumination
of the medium. It is a most remarkable fact that the monodirectional illumination problem can be reduced to the isotropic illumination problem, while the converse is obvious.
This is accomplished through the decomposition formula which
gives the desired functions as simple multiplications and ad-
ditions of b and h functions. Another important point to note
is that these b and h functions can be easily computed. We
present a number of selected tables and graphs of these functions,and we show how to find, by using a small hand or desk
calculator, all of the interesting functions of the basic
problem.

In the third part of this chapter, we discuss the
solution of the integral equation for the problem of radiative transfer in an isotropic slab bounded by a Lambert's law
reflector. It is shown that, for a kernel of the type

$$k(t,y) = \frac{1}{2} E_1(|t-y|) + A\, E_2(t)\, E_2(y) ,$$

where

$$E_1(r) = \int_0^1 e^{-r/z}\, dz/z ,$$

$$E_2(r) = \int_0^1 e^{-r/z}\, dz , \qquad\qquad r>0,$$

it is not necessary to compute the auxiliary reflection function r(v,z,x), which is a function of three arguments.
Rather, there are X and Y functions of fewer arguments which
suffice to complete the Cauchy system. This implies that
reductions may be possible for other kernels for which this
has not yet been done. Finally, we discuss inverse problems.

9.2   PHYSICAL DERIVATION OF CAUCHY SYSTEM FOR
           SOURCE FUNCTION

We consider the determination of the internal radiation field due to multiple scattering in a finite, homogeneous, isotropically scattering horizontal medium illuminated
uniformly at the top by parallel rays of radiation (monodirectional illumination). No radiation illuminates the
medium from below. Any radiation that falls on the lower
boundary from above is completely absorbed. The thickness
of the medium is x. The unit of length (optical length) is
chosen so that in a path of length Δ, the fraction $\Delta + o(\Delta)$
of the radiation propagating along that path is absorbed,
temporarily or otherwise. The albedo for single scattering,
λ, is the fraction of the absorbed energy which is re-emitted
when a single scattering event takes place, and $0 \leq \lambda \leq 1$.
The unit of energy is chosen so that the incident
radiation is π per unit normal area per unit time. Polar
angles, arccosine z, are measured from the inward directed
vertical, and $0 \leq z \leq 1$. The problem is to determine the
internal and external radiation fields.
The source function J(t,x,z) is the rate of production of scattered radiation per unit volume per unit solid
angle at optical altitude t, the thickness being x and

direction cosine of incident radiation being z. The source
function satisfies the Fredholm integral equation

$$J(t,x,z) = \frac{\lambda}{4} e^{-(x-t)/z}$$

$$+ \frac{\lambda}{2} \int_0^x J(t',x,z) \, E_1(|t-t'|) \, dt' \, , \qquad (9.1)$$

$$E_1(|s|) = \int_0^1 e^{-|s|/z} \frac{dz}{z} \, . \qquad (9.2)$$

The first term on the right-hand side of Eq. (9.1) arises
from the reduced incident radiation and the second from the
diffuse radiation field. As we have learned, this integral
equation is equivalent to a Cauchy system. We would prefer
to deal with the latter which provides us with a desirable
parameter study in both thickness and direction.

Consider the slab of thickness x and its source
function $J(t,x,z)$, Consider the slab of thickness x + Δ
and its source function $J(t,x + \Delta, z)$. Regard the slab of
thickness x + Δ as being formed by the addition of a thin
layer of thickness Δ to the top of the slab of thickness
x. We maintain the monodirectional illumination of the top
of the medium, the incident angle being arc cos z. Now consider the effect of the additional thin layer upon the source
function at the fixed altitude t. There is a dimunition due
to weakening of the incident radiation as it passes through
the thin layer. On the other hand, there is an increase due
to scattering at the top which leads to production of scattered radiation at altitude t. Other processes contribute
terms involving Δ^2 and higher order terms in Δ. This
leads to the equation

$$J(t,x+\Delta,z) = J(t,x,z) - \frac{\Delta}{z} J(t,x,z)$$

$$+ J(x,x,z) \frac{\Delta}{\pi} \int_0^1 J(t,x,z') \frac{2\pi dz'}{z'} \qquad (9.3)$$

$$+ o(\Delta) \; .$$

The function $o(\Delta)$ includes second and higher-order terms in Δ.

We can write an expansion for $J(t,x+\Delta,z)$ as

$$J(t,x+\Delta,z) = J(t,x,z) + \Delta \frac{\partial}{\partial x} J(t,x,z)$$

$$\qquad (9.4)$$

$$+ \Delta^2 \frac{\partial^2}{\partial x^2} J(t,x,z) + \dots \quad .$$

Equate the two expressions for $J(t,x+\Delta,z)$, and divide through by Δ. We obtain

$$\frac{\partial}{\partial x} J(t,x,z) + \Delta \frac{\partial^2}{\partial x^2} J(t,x,z) = - \frac{1}{z} J(t,x,z)$$

$$+ 2 J(x,x,z) \int_0^1 J(t,x,z') \frac{dz'}{z'} \qquad (9.5)$$

$$+ o(\Delta) \; .$$

In the limit as $\Delta \to 0$, this becomes

$$\frac{\partial}{\partial x} J(t,x,z) = - \frac{1}{z} J(t,x,z)$$

$$\qquad (9.6)$$

$$+ 2 J(x,x,z) \int_0^1 J(t,x,z') \frac{dz'}{z'} \; .$$

An equation is needed for the function $J(x,x,z)$, which is the source function at the top of the slab. To derive it, add a layer of thickness Δ to the bottom of the slab of thickness x. A relation between the two souce functions $J(x,x,z)$ and $J(x+\Delta,x+\Delta,z)$ is

$$J(x+\Delta,x+\Delta,z) = J(x,x,z)$$

$$+ J(0,x,z) \frac{\Delta}{\pi} \int_0^1 J(0,x,z') \frac{2\pi dz'}{z'} \qquad (9.7)$$

$$+ o(\Delta) ,$$

where $J(0,x,z)$ is the source function at the bottom, i.e., the side of the slab away from the illumination. In the limit as $\Delta \to 0$ this becomes

$$\frac{d}{dx} J(x,x,z) = 2 J(0,x,z) \int_0^1 J(0,x,z') \frac{dz'}{z'} . \qquad (9.8)$$

We readily obtain the equation for $J(0,x,z)$ by setting $t = 0$ in Eq. (9.6):

$$\frac{\partial}{\partial x} J(0,x,z) = - \frac{1}{z} J(0,x,z)$$

$$\qquad (9.9)$$

$$+ 2 J(x,x,z) \int_0^1 J(t,x,z') \frac{dz'}{z'} .$$

Our system of differential equations is now complete.

The initial conditions are determined by the consideration of a slab of thickness Δ. The source functions at "top" and "bottom" are given by the relations

$$J(\Delta,\Delta,z) \quad = J(0,\Delta,z)\ \Delta$$

$$= \pi\ \frac{\lambda}{4\pi}\ \Delta + o(\Delta)\ ,$$

(9.10)

which leads to the conditions

$$J(x,x,z)\big|_{x=0} = \frac{\lambda}{4}\ ,$$

(9.11)

$$J(0,x,z)\big|_{x=0} = \frac{\lambda}{4}\ ,$$

(9.12)

and further it is obvious that

$$J(t,x,z)\big|_{x=t} = J(x,x,z)\big|_{x=t}\ .$$

(9.13)

It is customary to normalize so that the initial conditions are unity. We introduce the functions X and Y through the formulas

$$J(x,x,z) = \frac{\lambda}{4}\ X(x,z)\ ,$$

(9.14)

$$J(0,x,z) = \frac{\lambda}{4}\ Y(x,z)\ .$$

(9.15)

These are functions of only two variables.

The complete initial value problem is expressed by the system of differential equations

$$\frac{\partial}{\partial x} J(t,x,z) = -\frac{1}{z} J(t,x,z)$$

$$+ \frac{\lambda}{2} X(x,z) \int_0^1 J(t,x,z') \frac{dz'}{z'} \ , \qquad (9.16)$$

$$\frac{\partial}{\partial x} X(x,z) = \frac{\lambda}{2} Y(x,z) \int_0^1 Y(x,z') \frac{dz'}{z'} \ , \qquad (9.17)$$

$$\frac{\partial}{\partial x} Y(x,z) = -\frac{1}{z} Y(x,z) + \frac{\lambda}{2} X(x,z)$$

$$\qquad\qquad\qquad\qquad\qquad\qquad\qquad (9.18)$$

$$\cdot \int_0^1 Y(x,z') \frac{dz'}{z'} \ ,$$

and the initial conditions

$$J(t,x,z)\big|_{x=t} = \frac{\lambda}{4} X(t,z) \ , \qquad (9.19)$$

$$X(0,z) = 1 \ , \qquad (9.20)$$

$$Y(0,z) = 1 \ , \qquad (9.21)$$

for $\ 0 < x \le 1, \quad 0 < z \le 1, \quad x \le t.$

9.3   COMPUTATIONAL METHOD

We use Gaussian quadrature for the evaluation of integrals,

$$\int_0^1 f(z') \, dz' \cong \sum_{i=1}^N f(z_i) \, w_i \ , \qquad (9.22)$$

where the abscissas $z_1, z_2, \ldots, z_N$ are the roots of the shifted Legendre polynomial $P_N(1-2z)$, and the weights $w_1, w_2, \ldots, w_N$ are the corresponding Christoffel numbers. For $N = 7$ the roots, weights, and angles arccos z_i are given in Table 9.1.

Table 9.1

Roots, Weights, and Angles for $N = 7$

i	Root z_i	Weight w_i	Angle, Arc cosine z_i (in degrees)
1	.025446046	.06474248	88. 541891
2	.12923441	.13985270	82. 574646
3	.29707742	.19091503	72. 717849
4	.50000000	.20897959	60. 000000
5	.70292258	.19091503	45. 338044
6	.87076559	.13985270	29. 452271
7	.97455396	.06474248	12. 953079

We introduce the functions of the one variable, x,

$$J_i(t,x) = J(t,x,z_i) ,$$

$$X_i(x) = X(x,z_i) , \qquad\qquad (9.23)$$

$$Y_i(x) = Y(x,z_i) , \qquad\qquad i=1,2,\ldots,N.$$

where t is a fixed parameter. These functions satisfy the system of ordinary differential equations

$$\frac{d}{dx} J_i = -\frac{1}{z_i} J_i + \frac{\lambda}{2} X_i \sum_{j=1}^{N} J_j \frac{w_j}{z_j} , \qquad (9.24)$$

$$\frac{d}{dx} X_i = \frac{\lambda}{2} Y_i \sum_{j=1}^{N} Y_j \frac{w_j}{z_j} , \qquad (9.25)$$

$$\frac{d}{dx} Y_i = -\frac{1}{z_i} Y_i + \frac{\lambda}{2} X_i \sum_{j=1}^{N} Y_j \frac{w_j}{z_j} , \qquad (9.26)$$

$$i = 1, 2, \ldots, N,$$

and the initial conditions

$$J_i(t,t) = \frac{\lambda}{4} X_i(t) , \qquad (9.27)$$

$$X_i(0) = 1 , \qquad (9.28)$$

$$Y_i(0) = 1 , \qquad\qquad i = 1, 2, \ldots, N. \quad (9.29)$$

The procedure is to use Eqs. (9.28) and (9.29) for initial conditions on X_1, X_2, ..., X_N, Y_1, Y_2, ..., Y_N, and integrate Eqs. (9.25) and (9.26) until x has attained the value of the altitude t at which J_1, J_2, ..., J_N are to be evaluated. The conditions on the Js are given by Eq. (9.27) in terms of the known X s. The integration of the enlarged system of 3N differential equations (9.24) – (9.26) proceeds until x has the largest value for which the study is desired. The result is the tabulation of the J functions evaluated at the fixed altitude t for N incident directions for a range of thicknesses. In addition one has the X and Y functions for N incident directions for the same range of thicknesses.

If source functions are desired at another altitude t_i, then N additional differential equations like (9.24) are adjoined when x reaches the value t_1, and N initial conditions are known of the type of Eq. (9.27), and so on for additional altitudes. If a source function $J(t,x,v)$ is desired for an incident direction whose cosine is v, where v is not one of the roots $z_1, z_2, \ldots, z_N$, then one could adjoin the three equations

$$\frac{d}{dx} J(t,x,v) = -\frac{1}{v} J(t,x,v)$$

$$+ \frac{\lambda}{2} X(x,v) \sum_{j=1}^{N} J_j \frac{w_j}{z_j} , \qquad (9.30)$$

$$\frac{d}{dx} X(x,v) = -\frac{\lambda}{2} Y(x,v) \sum_{j=1}^{N} Y_j \frac{w_j}{z_j} , \qquad (9.31)$$

$$\frac{d}{dx} Y(x,v) = -\frac{1}{v} Y(x,v) + \frac{\lambda}{2} X(x,v) \sum_{j=1}^{N} Y_j \frac{w_j}{z_j} ,$$

$$(9.32)$$

with suitable initial conditions.

9.4   NUMERICAL RESULTS

We use $N = 7$ for the quadrature formula and a step size of 0.005 for an Adams-Moulton fourth-order numerical integration method, for we are primarily interested in high accuracy and secondarily in computing times. Variation of both N and the step size, and comparison with the limited results which are available in the literature assure us that

our computations are at least as accurate as any other,
and extend over a wider range of thicknesses. No numerical
difficulty is encountered for the special case $\lambda = 1$.

Some X and Y and source functions are found in
Tables 9.2 and 9.3, and in Figures 9.1 through 9.8. In
Table 9.2, X and Y functions are evaluated at various
thicknesses and for the seven angles given in Table 9.1.
There are two rows of values for every thickness. The upper
row is for X and the lower one is for Y. In Table 9.3
are source functions for $\lambda = 0.3$, x = 5, and $\lambda = 0.5$,
x = 10, for seven incident directions. The source functions
are tabulated at depths x - t = 0, 0.5, 1.0, ..., 10.0.

The X and Y functions are plotted as functions
of angles, with thickness as a parameter and λ being fixed
at 0.4, (Figures 9.1 and 9.2) and 1.0 (Figures 9.3 and 9.4).
Source functions are plotted as functions of depth for
selected incident directions and for the four cases, $\lambda = 0.2$,
0.4, 0.99 and 1.0 in Figures 9.5 through 9.8. The case
$\lambda = 1.0$ is very special, and differs markedly from those
cases for which $\lambda < 1.0$.

9.5   THE b AND h FUNCTIONS FOR ISOTROPIC ILLUMINATION

Consider now the problem of multiple scattering of
radiation in a horizontal plane-parallel, homogeneous, isotropic medium which is illuminated at the top by isotropic
radiation of one unit of energy per unit of horizontal area
per unit of solid angle per unit of time. Let the source
function at altitude t for a slab of thickness x be denoted $\Phi(t,x)$. The source function satisfies the Fredholm
integral equation

Table 9.2

X AND Y FUNCTIONS,   ALBEDO = 1.0000

THICKNESS	1	2	3	4	5	6	7
0.	1.00000E 00	1.00000E 00	1.00000E 00	1.00000E 00	1.00000E 00	1.00000E 00	1.00000E 00
	1.00000E 00	1.00000E 00	1.00000E 00	1.00000E 00	1.00000E 00	1.00000E 00	1.00000E 00
0.1000	1.05294E 00	1.12251E 00	1.14461E 00	1.15277E 00	1.15643E 00	1.15822E 00	1.15902E 00
	5.22771E-02	5.70326E-01	8.51605E-01	9.66939E-01	1.02049E 00	1.04699E 00	1.05904
0.2000	1.05625E 00	1.16641E 00	1.22118E 00	1.24456E 00	1.25563E 00	1.26116E 00	1.26369E 00
	2.54532E-02	3.40591E-01	7.06569E-01	8.98214E-01	9.95467E-01	1.04555E 00	1.06876E 00
0.3000	1.05814E 00	1.18911E 00	1.27424E 00	1.31538E 00	1.33586E 00	1.34633E 00	1.35116E 00
	2.12533E-02	2.23725E-01	5.91162E-01	8.30273E-01	9.62275E-01.	1.03290E 00	1.06621E 00
0.4000	1.05957E 00	1.20274E 00	1.31323E 00	1.37252E 00	1.40344E 00	1.41960E 00	1.42714E 00
	1.87356E-02	1.62374E-01	5.00625E-01	7.66827E-01	9.25945E-01	1.01426E 00	1.05663E 00
0.5000	1.06071E 00	1.21200E 00	1.34306E 00	1.41992E 00	1.46174E 00	1.48406E 00	1.49459E 00
	1.69278E-02	1.28590E-01	4.29842E-01	7.09039E-01	8.88883E-01	9.92276E-01	1.04273E 00
0.6000	1.06166E 00	1.21893E 00	1.36663E 00	1.46003E 00	1.51287E 00	1.54165E 00	1.55535E 00
	1.55453E-02	1.08768E-01	3.74430E-01	6.57095E-01	8.52384E-01	9.68519E-01	1.02612E 00
0.7000	1.06248E 00	1.22448E 00	1.38576E 00	1.49448E 00	1.55826E 00	1.59364E 00	1.61064E 00
	1.44380E-02	9.62122E-02	3.30839E-01	6.10718E-01	8.17129E-01	9.43924E-01	1.00782E 00
0.8000	1.06319E 00	1.22914E 00	1.40166E 00	1.52443E 00	1.59889E 00	1.64094E 00	1.66133E 00
	1.35202E-02	8.75825E-02	2.96285E-01	5.69436E-01	7.83457E-01	9.19073E-01	9.88463E-01
0.9000	1.06381E 00	1.23316E 00	1.41513E 00	1.55073E 00	1.63554E 00	1.68424E 00	1.70806E 00
	1.27398E-02	8.11830E-02	2.68628E-01	5.32715E-01	7.51518E-01	8.94338E-01	9.68501E-01
1.0000	1.06437E 00	1.23671E 00	1.42674E 00	1.57403E 00	1.66878E 00	1.72406E 00	1.75132E 00
	1.20637E-02	7.61310E-02	2.46243E-01	5.00032E-01	7.21352E-01	8.69961E-01	9.48239E-01
1.1000	1.06488E 00	1.23988E 00	1.43689E 00	1.59483E 00	1.69908E 00	1.76083E 00	1.79153E 00
	1.14696E-02	7.19512E-02	2.27902E-01	4.70898E-01	6.92939E-01	8.46104E-01	9.27898E-01
1.2000	1.06534E 00	1.24274E 00	1.44588E 00	1.61352E 00	1.72683E 00	1.79491E 00	1.82900E 00
	1.09414E-02	6.83759E-02	2.12680E-01	4.44872E-01	6.66223E-01	8.22870E-01	9.07642E-01
1.3000	1.06575E 00	1.24535E 00	1.45392E 00	1.63042E 00	1.75233E 00	1.82658E 00	1.86403E 00
	1.04675E-02	6.52461E-02	1.99880E-01	4.21562E-01	6.41129E-01	8.00329E-01	8.87592E-01
1.4000	1.06614E 00	1.24774E 00	1.46118E 00	1.64578E 00	1.77586E 00	1.85609E 00	1.89684E 00
	1.00391E-02	6.24613E-02	1.88975E-01	4.00626E-01	6.17570E-01	7.78522E-01	8.67837E-01
1.5000	1.06649E 00	1.24993E 00	1.46780E 00	1.65983E 00	1.79764E 00	1.88367E 00	1.92766E 00
	9.64924E-03	5.99542E-02	1.79566E-01	3.81764E-01	5.95456E-01	7.57470E-01	8.48445E-01
1.6000	1.06682E 00	1.25197E 00	1.47386E 00	1.67272E 00	1.81785E 00	1.90949E 00	1.95664E 00
	9.29234E-03	5.76767E-02	1.71351E-01	3.64715E-01	5.74695E-01	7.37181E-01	8.29465E-01
1.7000	1.06712E 00	1.25385E 00	1.47944E 00	1.68461E 00	1.83666E 00	1.93371E 00	1.98396E 00
	8.96400E-03	5.55933E-02	1.64098E-01	3.49252E-01	5.55198E-01	7.17651E-01	8.10930E-01
1.8000	1.06740E 00	1.25561E 00	1.48461E 00	1.69561E 00	1.85422E 00	1.95649E 00	2.00975E 00
	8.66059E-03	5.36763E-02	1.57628E-01	3.35181E-01	5.36876E-01	6.98871E-01	7.92866E-01
1.9000	1.06767E 00	1.25725E 00	1.48942E 00	1.70583E 00	1.87065E 00	1.97794E 00	2.03413E 00
	8.37909E-03	5.19039E-02	1.51803E-01	3.22331E-01	5.19648E-01	6.80824E-01	7.75286E-01
2.0000	1.06792E 00	1.25879E 00	1.49391E 00	1.71535E 00	1.88604E 00	1.99817E 00	2.05721E 00
	8.11700E-03	5.02582E-02	1.46515E-01	3.10557E-01	5.03434E-01	6.63491E-01	7.58200E-01
2.1000	1.06815E 00	1.26023E 00	1.49812E 00	1.72426E 00	1.90051E 00	2.01729E 00	2.07909E 00
	7.87219E-03	4.87246E-02	1.41679E-01	2.99732E-01	4.88162E-01	6.46848E-01	7.41610E-01
2.2000	1.06837E 00	1.26159E 00	1.50207E 00	1.73260E 00	1.91413E 00	2.03538E 00	2.09987E 00
	7.64285E-03	4.72907E-02	1.37228E-01	2.89745E-01	4.73763E-01	6.30872E-01	7.25515E-01
2.3000	1.06858E 00	1.26287E 00	1.50578E 00	1.74044E 00	1.92698E 00	2.05253E 00	2.11961E 00
	7.42746E-03	4.59460E-02	1.33109E-01	2.80502E-01	4.60173E-01	6.15536E-01	7.09912E-01
2.4000	1.06877E 00	1.26408E 00	1.50929E 00	1.74783E 00	1.93911E 00	2.06879E 00	2.13840E 00
	7.22464E-03	4.46818E-02	1.29277E-01	2.71921E-01	4.47333E-01	6.00815E-01	6.94792E-01
2.5000	1.06896E 00	1.26523E 00	1.51261E 00	1.75480E 00	1.95060E 00	2.08425E 00	2.15629E 00
	7.03328E-03	4.34903E-02	1.25698E-01	2.63930E-01	4.35188E-01	5.86683E-01	6.80147E-01

Table 9.3

Source Functions

ALBEDO =0.3000   THICKNESS = 5.0000

DEPTH	J(1)	J(2)	J(3)	J(4)	J(5)	J(6)	J(7)
0.	7.61188E-02	7.85250E-02	8.06983E-02	8.23169E-02	8.34111E-02	8.40876E-02	8.43438E-02
0.5000	2.29722E-02	3.15939E-03	1.87018E-02	3.56806E-02	4.73932E-02	5.43857E-02	5.78693E-02
1.0000	1.00581E-04	6.69796E-04	4.80078E-03	1.49725E-02	2.54289E-02	3.29626E-02	3.70651E-02
1.5000	5.00890E-05	3.08519E-04	1.52827E-03	6.38839E-03	1.35119E-02	1.96715E-02	2.33380E-02
2.0000	2.63642E-05	1.59920E-04	6.13041E-04	2.80412E-03	7.16287E-03	1.16484E-02	1.45620E-02
2.5000	1.43221E-05	8.62229E-05	2.90849E-04	1.27523E-03	3.80027E-03	6.86498E-03	9.03321E-03
3.0000	7.93769E-06	4.75588E-05	1.51445E-04	6.03059E-04	2.02130E-03	4.03264E-03	5.57929E-03
3.5000	4.45810E-06	2.66215E-05	8.25662E-05	2.96532E-04	1.07853E-03	2.36192E-03	3.43222E-03
4.0000	2.52264E-06	1.50266E-05	4.59877E-05	1.50915E-04	5.76697E-04	1.37722E-03	2.09949E-03
4.5000	1.42243E-06	8.45682E-06	2.56823E-05	7.84580E-05	3.06665E-04	7.92948E-04	1.26632E-03
5.0000	7.46246E-07	4.42999E-06	1.33845E-05	3.87322E-05	1.51708E-04	4.21824E-04	7.04879E-04

ALBEDO =0.5000   THICKNESS = 10.0000

DEPTH	J(1)	J(2)	J(3)	J(4)	J(5)	J(6)	J(7)
0.	1.28289E-01	1.35744E-01	1.42888E-01	1.48467E-01	1.52368E-01	1.54834E-01	1.56116E-01
0.5000	8.14272E-04	8.15287E-03	3.94372E-02	7.35127E-02	9.75279E-02	1.12145E-01	1.19519E-01
1.0000	3.97152E-04	2.56262E-03	1.26258E-02	3.45147E-02	5.67113E-02	7.27658E-02	8.15519E-02
1.5000	2.15513E-04	1.33053E-03	5.13706E-03	1.66124E-02	3.24595E-02	4.59978E-02	5.40423E-02
2.0000	1.22148E-04	7.46070E-04	2.51545E-03	8.29603E-03	1.85097E-02	2.87006E-02	3.52611E-02
2.5000	7.08999E-05	4.30705E-04	1.36664E-03	4.30894E-03	1.05674E-02	1.77707E-02	2.27811E-02
3.0000	4.17638E-05	2.52855E-04	7.81141E-04	2.32154E-03	6.05482E-03	1.09491E-02	1.46169E-02
3.5000	2.48466E-05	1.50088E-04	4.57780E-04	1.29049E-03	3.48619E-03	6.72381E-03	9.33044E-03
4.0000	1.48878E-05	8.97824E-05	2.71949E-04	7.35541E-04	2.01838E-03	4.11977E-03	5.93218E-03
4.5000	8.96842E-06	5.40177E-05	1.62914E-04	4.27372E-04	1.17535E-03	2.52035E-03	3.75956E-03
5.0000	5.42491E-06	3.26433E-05	9.81567E-05	2.51905E-04	6.86408E-04	1.54028E-03	2.37641E-03
5.5000	3.29218E-06	1.97949E-05	5.93909E-05	1.50058E-04	4.05466E-04	9.40701E-04	1.49884E-03
6.0000	2.00314E-06	1.20369E-05	3.60525E-05	9.00840E-05	2.40082E-04	5.74293E-04	9.43578E-04
6.5000	1.22140E-06	7.33567E-06	2.19416E-05	5.43893E-05	1.42852E-04	3.50533E-04	5.93051E-04
7.0000	7.45996E-07	4.47853E-06	1.33807E-05	3.29760E-05	8.53744E-05	2.13394E-04	3.72185E-04
7.5000	4.56188E-07	2.73941E-06	8.17201E-06	2.00527E-05	5.12175E-05	1.30545E-04	2.33220E-04
8.0000	2.79093E-07	1.67441E-06	4.99419E-06	1.22149E-05	3.08141E-05	7.96128E-05	1.45858E-04
8.5000	1.70530E-07	1.02282E-06	3.04872E-06	7.43779E-06	1.85559E-05	4.84526E-05	9.09100E-05
9.0000	1.03549E-07	6.20941E-07	1.84979E-06	4.50390E-06	1.11273E-05	2.92899E-05	5.61994E-05
9.5000	6.14347E-08	3.68328E-07	1.09673E-06	2.66612E-06	6.53155E-06	1.72943E-05	3.38817E-05
10.0000	3.17138E-08	1.90106E-07	5.65825E-07	1.37372E-06	3.34080E-06	8.88277E-06	1.77477E-05

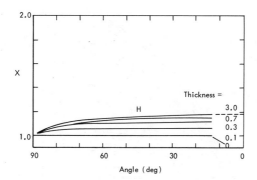

Figure 9.1   X   Functions   for $\lambda = 0.4$

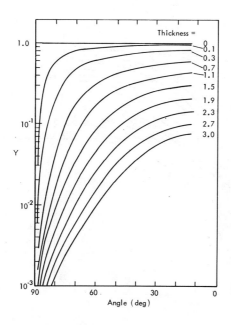

Figure 9.2   Y   Functions   for $\lambda = 0.4$

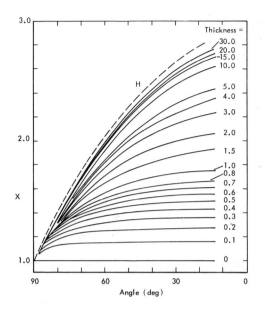

Figure 9.3   X   Functions   for λ = 1.0

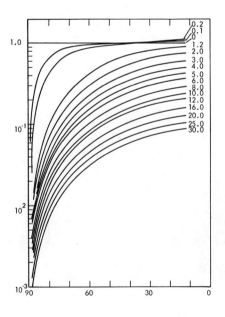

Figure 9.4   Y   Functions   for λ = 1.0

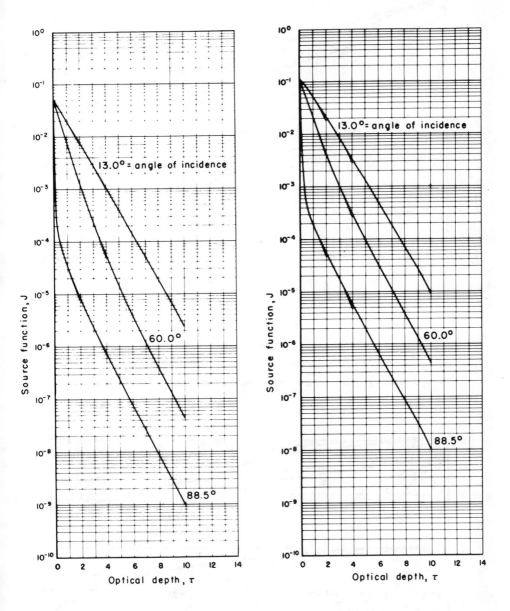

Figure 9.5   Several Source
             Functions for
             $\lambda = 0.2$

Figure 9.6   Several Source
             Functions for
             $\lambda = 0.4$

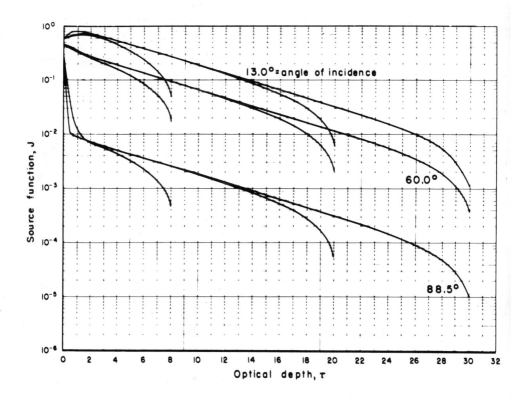

Figure 9.7   Several Source Functions for $\lambda = 0.99$

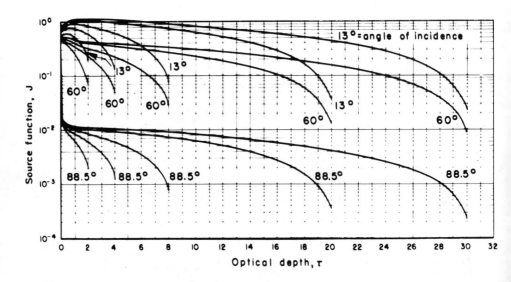

Figure 9.8   Several Source Functions for $\lambda = 1.0$

$$\Phi(t,x) = \frac{\lambda}{2} E_1(x-t)$$

$$+ \frac{\lambda}{2} \int_0^x \Phi(t',x) \, E_1(|t-t'|) \, dt' \, . \tag{9.33}$$

One realizes immediately upon comparison of this equation
for Φ and the corresponding integral equation for J that
the two source functions are related through the formula

$$\Phi(t,x) = 2 \int_0^1 J(t,x,z) \, dz/z \, . \tag{9.34}$$

It is natural that the solution for the isotropic source
problem be an integral over all directions of the solutions
for the monodirectional source problems. We will soon dis-
cover that a converse relation exists, and this is certainly
not obvious.

Let us introduce the function $b(t,v,x)$ as the internal intensity of the total radiation field at altitude t
in the direction whose cosine is v with respect to the upward directed vertical due to the described isotropic sources
at the top $(-1 \leq v \leq 1)$. The function $h(t,v,x)$ is defined as
the internal intensity at altitude t in the direction whose
cosine is v with respect to the upward directed vertical
when the isotropic sources are located at the bottom of the
slab. The b and h functions are related through the
formula

$$h(t,v,x) = b(x-t, -v, x) \tag{9.35}$$

which corresponds to "flipping the slab over."

Physically, and mathematically, the source functions and intensities are related through the equations

$$\Phi(t,x) = \frac{\lambda}{2} \int_0^1 b(t,v,x) \, dv \, ,$$
$$(9.36)$$

$$b(t,v,x) = \frac{1}{v} \int_0^t e^{-(t-y)/v} \, \Phi(y,x) \, dy \, ,$$
$$(9.37)$$
$$v > 0 \, ,$$

$$b(t,v,x)$$

$$= -\frac{1}{v} \left[e^{-(t-y)/v} + \int_t^x e^{-(t-y)/v} \, \Phi(y,x) \, dy \right] \, ,$$
$$v < 0 \, , \quad (9.38)$$

The source function for the isotropic sources at the bottom is $\Phi(x-t,x)$ at altitude t, and equations corresponding to Eqs. (9.36) - (9.38) exist for h and Φ. The b, h, and Φ functions are very interesting from both the physical and mathematical points of view.

Let us imbed the functions b and h in a physical manner. The equation for h is the easier equation to obtain. See Fig. 9.9. Add a layer of thickness Δ to the top of the slab of thickness x. This does not disturb the illumination on the slab. The function h for the enlarged slab is expressible as

$$h(t,v,x+\Delta) = h(t,v,x) + \Phi(0,x)\Delta \, b(t,v,x) + o(\Delta) \, ,$$
$$(9.39)$$

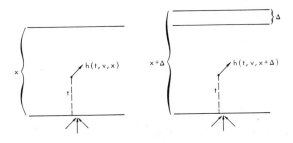

Figure 9.9   Imbedding Diagrams for the Function h

where the second term on the right is the intensity at
altitude t due to scattering at the top, and o (Δ) takes
higher order processes into account. By letting Δ → 0 ,
we obtain the differential equation,

$$h_x(t,v,x) = \Phi(0,x)\, b(t,v,x) \, ,\hspace{3em} (9.40)$$

where the subscript x denotes partial differentiation with
respect to x .

    For the imbedding of the b function, since we do
not wish to change the illumination on the slab of thickness
x + Δ, we construct the slab of thickness x + Δ by addition
of the thin layer to the bottom of the slab of thickness x.
See Fig. 9.10. This has two first order effects: the added
scattering in this thin slab ($\Phi(0,x)\Delta$) gives rise to intensity
h(t,v,x), and there is a change in b due to the change in

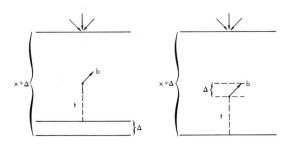

Figure 9.10   Imbedding Diagrams for the Function b

altitude from t to t + Δ . This latter change increases
b by the amount Δv^{-1} [b(t,v,x) - Φ(t,x)] . The difference
equation is

$$b(t,v,x+\Delta) = b(t,v,x) + \Phi(0,x) \Delta h(t,v,x)$$
$$+ \frac{\Delta}{v} [b(t,v,x) - \Phi(t,x)] + o(\Delta) ,$$

(9.41)

and the differential equation is

$$b_x(t,v,x) = \Phi(0,x) h(t,v,x)$$
$$+ v^{-1} [b(t,v,x) - \Phi(t,x)] .$$

(9.42)

The initial conditions when x = t are easy to find for the
negative direction cosines:

$$b(t,v,t) = -v^{-1} ,$$ $-1 \leq v < 0$,   (9.43)

$$h(t,v,t) = 0 ,$$ $-1 \leq v < 0$.   (9.44)

The first equation is written from knowledge of the intensity of the illumination, and the second from the fact that no radiation enters from the unilluminated boundary.

The differential equations (9.40) and (9.42) and the initial conditions in Eqs. (9.43) and (9.44) determine b and h for negative cosines.   Then use of the formulas (which follow from (9.35)) ,

$$h(t,v,x) = b(x-t,-v,x) , \qquad\qquad (9.45)$$

$$b(t,v,x) = h(x-t,-v,x) , \qquad\qquad (9.46)$$

give b and h for positive cosines.   The system of equations needs to be adjoined to Eq. (9.34) for Φ in terms of J and the complete initial value problem for J as given earlier.

9.6   NUMERICAL RESULTS FOR b AND h FUNCTIONS

Extensive calculations are performed to evaluate b and h functions for slabs of various thicknesses and albedos, and to examine the intensities at various altitudes and in various directions.   Some of the results are found in Figures 9.11 and 9.12, and in Tables 9.4 through 9.6.

In Figure 9.11, b functions are plotted for the case $\lambda = 0.5$, $x = 2$ as functions of angles for various altitudes.   Note that while b is singular at the top, it is well behaved everywhere else.   The case $\lambda = 1.0$. $x = 5$ is shown in Figure 9.12.   In Table 9.4, there are two rows of values for each "B" and "H" , the upper row for positive

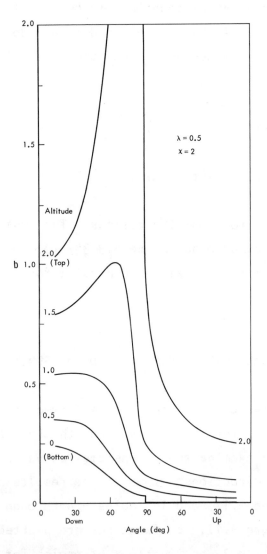

Figure 9.11   Some Functions   b   for $\lambda = 0.5$, $x = 2$

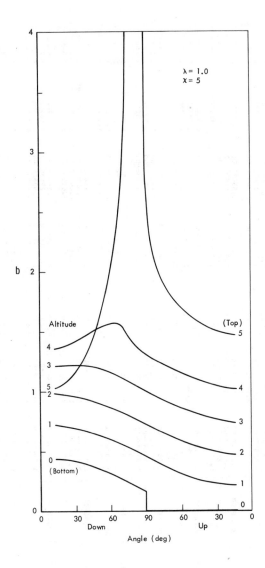

Figure 9.12   Some Functions   b   for $\lambda = 1.0$, $x = 5$

Table 9.4

Functions b, h, and Φ for λ = 1.0, x = 1.0

ALBEDO =1.0000 THICKNESS = 1.0000

DEPTH	PHI		1	2	3	4	5	6	7
0.	INFINITE	B	2.5297E 00	1.8316E 00	1.4365E 00	1.1481E 00	9.5143E-01	8.3152E-01	7.7094E-01
			3.9299E-01	7.7379E 00	3.3661E 00	2.0000E 00	1.4226E 00	1.1484E 00	1.0261E 00
		H	4.7409E-01	5.8909E-01	8.2889E-01	1.0001E 00	1.0262E 00	9.9908E-01	9.7300E-01
			0.	0.	0.	0.	0.	0.	0.
0.10	1.6703E 00	B	1.5884E 00	1.3801E 00	1.1439E 00	9.2585E-01	7.6892E-01	6.7203E-01	6.2292E-01
			2.5581E 00	4.6714E 00	3.0020E 00	2.0194E 00	1.5142E 00	1.2534E 00	1.1326E 00
		H	5.8559E-01	6.9738E-01	9.6025E-01	1.1103E 00	1.1063E 00	1.0595E 00	1.0238E 00
			5.2398E-01	2.7551E-01	1.4497E-01	9.1701E-02	6.7009E-02	5.4794E-02	4.9242E-02
0.20	1.3985E 00	B	1.3502E 00	1.1970E 00	9.9050E-01	7.9336E-01	6.5365E-01	5.6856E-01	5.2576E-01
			1.4733E 00	2.9644E 00	2.5762E 00	1.9280E 00	1.5145E 00	1.2820E 00	1.1701E 00
		H	6.8520E-01	8.0657E-01	1.1014E 00	1.2214E 00	1.1823E 00	1.1143E 00	1.0686E 00
			6.3466E-01	4.5932E-01	2.7787E-01	1.8601E-01	1.3921E-01	1.1517E-01	1.0404E-01
0.30	1.2222E 00	B	1.1852E 00	1.0566E 00	8.6224E-01	6.7920E-01	5.5359E-01	4.7860E-01	4.4128E-01
			1.2644E 00	2.0650E 00	2.2118E 00	1.8147E 00	1.4866E 00	1.2845E 00	1.1832E 00
		H	7.8405E-01	9.2744E-01	1.2594E 00	1.3352E 00	1.2548E 00	1.1636E 00	1.1075E 00
			7.3355E-01	5.9742E-01	4.0263E-01	2.8118E-01	2.1497E-01	1.7974E-01	1.6316E-01
0.40	1.0838E 00	B	1.0528E 00	9.3587E-01	7.4522E-01	5.7428E-01	4.6205E-01	3.9664E-01	3.6449E-01
			1.1174E 00	1.5677E 00	1.9074E 00	1.6940E 00	1.4419E 00	1.2698E 00	1.1799E 00
		H	8.8594E-01	1.0737E 00	1.4409E 00	1.4521E 00	1.3231E 00	1.2070E 00	1.1399E 00
			8.3343E-01	7.1469E-01	5.1949E-01	3.7710E-01	2.9386E-01	2.4807E-01	2.2619E-01
0.50	9.6588E-01	B	9.3822E-01	8.2460E-01	6.3250E-01	4.7436E-01	3.7598E-01	3.2021E-01	2.9315E-01
			9.9500E-01	1.2705E 00	1.6539E 00	1.5721E 00	1.3862E 00	1.2430E 00	1.1646E 00
		H	9.9500E-01	1.2705E 00	1.6539E 00	1.5721E 00	1.3862E 00	1.2430E 00	1.1646E 00
			9.3822E-01	8.2460E-01	6.3250E-01	4.7436E-01	3.7598E-01	3.2021E-01	2.9315E-01
0.60	8.5930E-01	B	8.3343E-01	7.1469E-01	5.1949E-01	3.7710E-01	2.9386E-01	2.4807E-01	2.2619E-01
			8.8594E-01	1.0737E 00	1.4409E 00	1.4521E 00	1.3231E 00	1.2070E 00	1.1399E 00
		H	1.1174E 00	1.5677E 00	1.9074E 00	1.6940E 00	1.4419E 00	1.2698E 00	1.1799E 00
			1.0528E 00	9.3587E-01	7.4522E-01	5.7428E-01	4.6205E-01	3.9664E-01	3.6449E-01
0.70	7.5867E-01	B	7.3355E-01	5.9742E-01	4.0263E-01	2.8118E-01	2.1497E-01	1.7974E-01	1.6316E-01
			7.8405E-01	9.2744E-01	1.2594E 00	1.3352E 00	1.2548E 00	1.1636E 00	1.1075E 00
		H	1.2644E 00	2.0650E 00	2.2118E 00	1.8147E 00	1.4866E 00	1.2845E 00	1.1832E 00
			1.1852E 00	1.0566E 00	8.6224E-01	6.7920E-01	5.5359E-01	4.7860E-01	4.4128E-01
0.80	6.6013E-01	B	6.3466E-01	4.5932E-01	2.7787E-01	1.8601E-01	1.3921E-01	1.1517E-01	1.0404E-01
			6.8520E-01	8.0657E-01	1.1014E 00	1.2214E 00	1.1823E 00	1.1143E 00	1.0686E 00
		H	1.4733E 00	2.9644E 00	2.5762E 00	1.9280E 00	1.5155E 00	1.2820E 00	1.1701E 00
			1.3502E 00	1.1970E 00	9.9050E-01	7.9336E-01	6.5365E-01	5.6856E-01	5.2576E-01
0.90	5.5970E-01	B	5.2398E-01	2.7551E-01	1.4497E-01	9.1701E-02	6.7009E-02	5.4794E-02	4.9242E-02
			5.8559E-01	6.9738E-01	9.6025E-01	1.1103E 00	1.1063E 00	1.0595E 00	1.0238E 00
		H	2.5581E 00	4.6714E 00	3.0020E 00	2.0194E 00	1.5142E 00	1.2534E 00	1.1326E 00
			1.5884E 00	1.3801E 00	1.1439E 00	9.2585E-01	7.6892E-01	6.7203E-01	6.2292E-01
1.00	4.3948E-01	B	0.	0.	0.	0.	0.	0.	0.
			4.7409E-01	5.8909E-01	8.2889E-01	1.0001E 00	1.0262E 00	9.9908E-01	9.7300E-01
		H	3.9299E-01	7.7379E 00	3.3661E 00	2.0000E 00	1.4226E 00	1.1484E 00	1.0261E 00
			2.5297E 00	1.8316E 00	1.4365E 00	1.1481E 00	9.5143E-01	8.3152E-01	7.7094E-01

Table 9.5

Functions b, h, and Φ for λ = 1.0, x = 10.0

ALBEDO =1.0000   THICKNESS = 10.0000

DEPTH	PHI		1	2	3	4	5	6	7
0.	INFINITE	B	2.8907E 00	2.2722E 00	2.0009E 00	1.8493E 00	1.7538E 00	1.6933E 00	1.6607E 00
			3.9299E 01	7.7379E 00	3.3661E 00	2.0000E 00	1.4226E 00	1.1484E 00	1.0261E 00
		H	9.4209E-02	1.1457E-01	1.4333E-01	1.7624E-01	2.0832E-01	2.3454E-01	2.5065E-01
			0.	0.	0.	0.	0.	0.	0.
1.00	1.5096E 00	B	1.5040E 00	1.4828E 00	1.4515E 00	1.4162E 00	1.3824E 00	1.3551E 00	1.3384E 00
			1.5154E 00	1.5461E 00	1.6617E 00	1.6963E 00	1.6154E 00	1.5192E 00	1.4574E 00
		H	2.6153E-01	2.7751E-01	3.0324E-01	3.4223E-01	3.6516E-01	3.9070E-01	4.0646E-01
			2.5365E-01	2.3741E-01	2.0893E-01	1.7564E-01	1.4900E-01	1.3172E-01	1.2275E-01
2.00	1.3277E 00	B	1.3236E 00	1.3070E 00	1.2806E 00	1.2490E 00	1.2177E 00	1.1919E 00	1.1759E 00
			1.3318E 00	1.3490E 00	1.3817E 00	1.4281E 00	1.4470E 00	1.4359E 00	1.4199E 00
		H	4.1455E-01	4.3033E-01	4.5584E-01	4.8666E-01	5.1748E-01	5.4295E-01	5.5862E-01
			4.0680E-01	3.9098E-01	3.6523E-01	3.3355E-01	3.0297E-01	2.7986E-01	2.6675E-01
3.00	1.1709E 00	B	1.1670E 00	1.1511E 00	1.1254E 00	1.0944E 00	1.0635E 00	1.0380E 00	1.0222E 00
			1.1748E 00	1.2172E 00	1.2712E 00	1.2518E 00	1.2828E 00	1.2980E 00	1.3016E 00
		H	5.6649E-01	5.8224E-01	6.0771E-01	6.3851E-01	6.6933E-01	6.9474E-01	7.1027E-01
			5.5876E-01	5.4300E-01	5.1750E-01	4.8652E-01	4.5562E-01	4.3081E-01	4.1605E-01
4.00	1.0181E 00	B	1.0142E 00	9.9843E-01	9.7292E-01	9.4209E-01	9.1128E-01	8.8581E-01	8.7011E-01
			1.0220E 00	1.0378E 00	1.0634E 00	1.0950E 00	1.1265E 00	1.1492E 00	1.1604E 00
		H	7.1823E-01	7.3398E-01	7.5945E-01	7.9027E-01	8.2116E-01	8.4644E-01	8.6158E-01
			7.1051E-01	6.9476E-01	6.6929E-01	6.3477E-01	6.0765E-01	5.8234E-01	5.6696E-01
5.00	8.6613E-01	B	8.6227E-01	8.4651E-01	8.2104E-01	7.9024E-01	7.5943E-01	7.3402E-01	7.1840E-01
			8.6999E-01	8.8575E-01	9.1126E-01	9.4221E-01	9.7331E-01	9.9804E-01	1.0121E 00
		H	8.6999E-01	8.8575E-01	9.1126E-01	9.4221E-01	9.7331E-01	9.9804E-01	1.0121E 00
			8.6227E-01	8.4651E-01	8.2104E-01	7.9024E-01	7.5943E-01	7.3402E-01	7.1840E-01
6.00	7.1437E-01	B	7.1051E-01	6.9476E-01	6.6929E-01	6.3847E-01	6.0763E-01	5.8234E-01	5.6696E-01
			7.1823E-01	7.3398E-01	7.5945E-01	7.9027E-01	8.2116E-01	8.4644E-01	8.6158E-01
		H	1.0220E 00	1.0378E 00	1.0634E 00	1.0950E 00	1.1265E 00	1.1492E 00	1.1604E 00
			1.0142E 00	9.9843E-01	9.7292E-01	9.4209E-01	9.1128E-01	8.8581E-01	8.7011E-01
7.00	5.6262E-01	B	5.5876E-01	5.4300E-01	5.1750E-01	4.8652E-01	4.5562E-01	4.3081E-01	4.1605E-01
			5.6649E-01	5.8224E-01	6.0771E-01	6.3851E-01	6.6933E-01	6.9474E-01	7.1027E-01
		H	1.1748E 00	1.2172E 00	1.2712E 00	1.2518E 00	1.2828E 00	1.2980E 00	1.3016E 00
			1.1670E 00	1.1511E 00	1.1254E 00	1.0944E 00	1.0635E 00	1.0380E 00	1.0222E 00
8.00	4.1067E-01	B	4.0680E-01	3.9098E-01	3.6523E-01	3.3355E-01	3.0297E-01	2.7986E-01	2.6675E-01
			4.1455E-01	4.3033E-01	4.5584E-01	4.8666E-01	5.1748E-01	5.4295E-01	5.5862E-01
		H	1.3318E 00	1.3490E 00	1.3817E 00	1.4281E 00	1.4470E 00	1.4359E 00	1.4199E 00
			1.3236E 00	1.3070E 00	1.2806E 00	1.2490E 00	1.2177E 00	1.1919E 00	1.1759E 00
9.00	2.5759E-01	B	2.5365E-01	2.3741E-01	2.0893E-01	1.7564E-01	1.4900E-01	1.3172E-01	1.2275E-01
			2.6153E-01	2.7751E-01	3.0324E-01	3.4223E-01	3.6516E-01	3.9070E-01	4.0646E-01
		H	1.5154E 00	1.5461E 00	1.6617E 00	1.6963E 00	1.6154E 00	1.5192E 00	1.4574E 00
			1.5040E 00	1.4828E 00	1.4515E 00	1.4162E 00	1.3824E 00	1.3551E 00	1.3384E 00
10.00	8.7559E-02	B	0.	0.	0.	0.	0.	0.	0.
			9.4209E-02	1.1457E-01	1.4333E-01	1.7624E-01	2.0832E-01	2.3454E-01	2.5065E-01
		H	3.9299E 01	7.7379E 00	3.3661E 00	2.0000E 00	1.4226E 00	1.1484E 00	1.0261E 00
			2.8907E 00	2.2722E 00	2.0009E 00	1.8493E 00	1.7538E 00	1.6933E 00	1.6607E 00

Table 9.6

Functions b, h, and Φ for λ = 0.5, x = 5.0

ALBEDO =0.5000   THICKNESS = 5.0000

DEPTH	PHI:		1	2	3	4	5	6	7
0.	INFINITE	B	1.0341E 00	6.6506E-01	4.8171E-01	3.7547E-01	3.1147E-01	2.7409E-01	2.5542E-01
		H	3.3299E 01	7.7379E 00	3.3661E 00	2.0000E 00	1.4226E 00	1.1484E 00	1.0261E 00
			1.3629E-03	1.6149E-03	2.1130E-03	3.2288E-03	6.2867E-03	1.1298E-02	1.5501E-02
			0.	0.	0.	0.	0.	0.	0.
0.50	2.3881E-01	B	2.2946E-01	1.9959E-01	1.6670E-01	1.3980E-01	1.2003E-01	1.0870E-01	1.0238E-01
		H	2.4943E-01	4.6474E-01	9.2905E-01	9.9029E-01	9.1060E-01	8.3178E-01	7.8557E-01
			2.6208E-03	2.9674E-03	3.7431E-03	5.6763E-03	1.0904E-02	1.8630E-02	2.4651E-02
			2.4742E-03	2.1838E-03	1.6848E-03	1.2673E-03	1.0059E-03	8.5734E-04	7.8516E-04
1.00	1.1766E-01	B	1.1396E-01	1.0130E-01	8.6221E-02	7.3303E-02	6.3832E-02	5.7697E-02	5.4466E-02
		H	1.2166E-01	1.4608E-01	2.9743E-01	4.6495E-01	5.2950E-01	5.3971E-01	5.3602E-01
			4.4930E-03	5.0574E-03	6.3557E-03	9.8642E-03	1.8807E-02	3.0522E-02	3.6957E-02
			4.2584E-03	3.8436E-03	3.2723E-03	2.7026E-03	2.2729E-03	2.0007E-03	1.8608E-03
1.50	6.4049E-02	B	6.2210E-02	5.5764E-02	4.7873E-02	4.0947E-02	3.5795E-02	3.2425E-02	3.0641E-02
		H	6.6014E-02	7.5841E-02	1.2101E-01	2.2378E-01	3.0307E-01	3.4117E-01	3.5520E-01
			7.5731E-03	8.5212E-03	1.0755E-02	1.7359E-02	3.2528E-02	4.9851E-02	6.1307E-02
			7.1821E-03	6.4971E-03	5.6161E-03	4.7819E-03	4.1302E-03	3.6988E-03	3.4710E-03
2.00	3.6355E-02	B	3.5362E-02	3.1842E-02	2.7466E-02	2.3572E-02	2.0645E-02	1.8717E-02	1.7691E-02
		H	3.7410E-02	4.2522E-02	5.9251E-02	1.1175E-01	1.7825E-01	2.1287E-01	2.3176E-01
			1.2771E-02	1.4394E-02	1.3855E-02	1.2593E-01	5.6520E-02	8.1198E-02	9.6062E-02
			1.2107E-02	1.0948E-02	9.4809E-03	8.1057E-03	7.1057E-03	6.4161E-03	6.0486E-03
2.50	2.1113E-02	B	2.0554E-02	1.8558E-02	1.6052E-02	1.3798E-02	1.2084E-02	1.0948E-02	1.0342E-02
		H	2.1705E-02	2.4540E-02	3.2135E-02	5.8037E-02	8.6558E-02	1.3180E-01	1.4973E-01
			2.1706E-02	2.4540E-02	3.2135E-02	5.0037E-02	9.8658E-02	1.3180E-01	1.4973E-01
			2.0554E-02	1.8558E-02	1.6052E-02	1.3798E-02	1.2084E-02	1.0948E-02	1.0342E-02
3.00	1.2430E-02	B	1.2107E-02	1.0948E-02	9.4809E-03	8.1401E-03	7.1057E-03	6.4161E-03	6.0488E-03
		H	1.2771E-02	1.4394E-02	1.8305E-02	3.1259E-02	6.5205E-02	8.1198E-02	9.6062E-02
			3.7410E-02	4.2522E-02	5.9252E-02	1.1175E-01	1.7282E-01	2.1287E-01	2.3176E-01
			3.5362E-02	3.1842E-02	2.7465E-02	2.3572E-02	2.0645E-02	1.8717E-02	1.7691E-02
3.50	7.3724E-03	B	7.1821E-03	6.4971E-03	5.6161E-03	4.8302E-03	4.1302E-03	3.6986E-03	3.4710E-03
		H	7.5731E-03	8.5212E-03	1.0755E-02	1.7359E-02	3.2528E-02	4.9851E-02	6.1307E-02
			6.6014E-02	7.5841E-02	1.2101E-01	2.2378E-01	3.0307E-01	3.4117E-01	3.5520E-01
			6.2210E-02	5.5764E-02	4.7873E-02	4.0947E-02	3.5795E-02	3.2425E-02	3.0641E-02
4.00	4.3727E-03	B	4.2584E-03	3.8436E-03	3.2223E-03	2.7026E-03	2.2729E-03	2.0007E-03	1.8608E-03
		H	4.4930E-03	5.0574E-03	6.3547E-03	9.8642E-03	1.8807E-02	3.0522E-02	3.6957E-02
			1.2166E-01	1.4608E-01	2.9743E-01	4.6495E-01	5.2950E-01	5.3971E-01	5.3602E-01
			1.1396E-01	1.0130E-01	8.6221E-02	7.3303E-02	6.3832E-02	5.7697E-02	5.4466E-02
4.50	2.5459E-03	B	2.4742E-03	2.1838E-03	1.6849E-03	1.2673E-03	1.0059E-03	8.5734E-04	7.8516E-04
		H	2.6208E-03	2.9674E-03	3.7431E-03	5.6742E-03	1.0904E-02	1.8630E-02	2.4651E-02
			2.4943E-01	4.6474E-01	9.2905E-01	9.9029E-01	9.1060E-01	8.3178E-01	7.8557E-01
			2.2946E-01	1.9959E-01	1.6670E-01	1.3980E-01	1.2003E-01	1.0870E-01	1.0238E-01
5.00	1.2940E-03	B	1.3629E-03	1.6149E-03	2.1130E-03	3.2288E-03	6.2867E-03	1.1298E-02	1.5501E-02
		H	0.	0.	0.	0.	0.	0.	0.
			3.3299E 01	7.7379E 00	3.3661E 00	2.0000E 00	1.4226E 00	1.1484E 00	1.0261E 00
			1.0341E 00	6.6506E-01	4.8171E-01	3.7547E-01	3.1147E-01	2.7409E-01	2.5542E-01

direction cosines and the lower for negative direction cosines. The numbers "1" through "7" refer to the seven
directions of propagation at the indicated depths. Also
tabulated is the Φ function, which is infinite at the top.
The table is given for the case $\lambda = 1.0$, $x = 1.0$. Tables
9.5 and 9.6 are for the cases $\lambda = 1.0$, $x = 10.0$ and $\lambda = 0.5$,
$x = 5.0$.

9.7   REDUCTION OF MONODIRECTIONAL SOURCE PROBLEM TO
         ISOTROPIC SOURCE PROBLEM

We will derive a decomposition formula which expresses all of the most relevant functions for the monodirectional illumination problem as simple algebraic operations on the b and h functions of the isotropic
illumination problem. The derivation requires two steps.
         Consider the internal intensity function $I(t,v,x,z)$
for the monodirectional illumination problem with incident
direction cosine z , slab thickness x , direction cosine
of propagation v $(-1 \leq v \leq +1)$, at altitude t . There are
two ways of expressing the change in I due to a change in
altitude. The first results from following a beam of radiation propagating from t to $t+\Delta$ with direction cosine v:

$$I(t+\Delta,v,x,z) = I(t,v,x,z) - \Delta v^{-1} I(t,v,x,z)$$

$$+ \Delta v^{-1} J(t,x,z) + o(\Delta) .$$

(9.47)

The loss term is due to absorption along the path, and the
gain is due to scattering into the direction of interest.
In the limit as $\Delta \to 0$, this becomes the transport equation,

$$v \, I_t(t,v,x,z) = -I(t,v,x,z) + J(t,x,z) \; . \qquad (9.48)$$

We can achieve the same increment in altitude by removing a slab of thickness Δ from the top and adding it to the bottom, the overall thickness remaining unchanged:

$$I(t+\Delta,v,x,z) = I(t,v,x,z) + \Delta z^{-1} \, I(t,v,x,z)$$
$$- J(x,x,z) \; \Delta \; b(t,v,x) \qquad (9.49)$$
$$+ J(0,x,z) \; \Delta \; h(t,v,x) + o \, (\Delta) \; .$$

The second term on the right is due to removal of the thin layer from the top, thus restoring the energy that would have been absorbed; the third term is due to the loss of scattering at the top by removal of the top layer, and the next term is due to gain in scattering at the bottom. The limit of this equation as $\Delta \rightarrow 0$ is the differential equation,

$$I_t(t,v,x,z) = z^{-1} \, I(t,v,x,z) - J(x,x,z) \, b(t,v,x)$$
$$\qquad (9.50)$$
$$+ J(0,x,z) \, h(t,v,x) \; .$$

Elimination of the partial derivatives in Eqs. (9.48) and (9.50) leads to one form of the desired decomposition equation,

$$(z^{-1} + v^{-1}) \, I(t,v,x,z) =$$
$$= v^{-1} \, J(t,x,z) \qquad (9.51)$$
$$+ J(x,x,z) \, b(t,v,x) - J(0,x,z) \, h(t,v,x) \; .$$

In terms of X and Y functions, this equation becomes

$$(z^{-1} + v^{-1}) \; I(t,v,x,z)$$

$$= v^{-1} \; J(t,x,z) \qquad\qquad (9.52)$$

$$+ \frac{\lambda}{4} \; X(x,z) \; b(t,v,x) - \frac{\lambda}{4} \; Y(x,z) \; h(t,v,x) \; .$$

Note the fact that I , which is a function of four variables,
is expressed in terms of functions of two and three variables,
an important reduction. We can go further and express the
X , Y and J functions in terms of b and h functions,
and consequently I is expressed entirely in terms of b
and h .

From the alternate physical meanings of X(x,z) and
Y(x,z) as probabilities of emergence with direction cosine
z at the top and the bottom, respectively, we can write those
probabilities as probabilities of direct emergence or
emergence after multiple scatterings. The equation for X(x,z)
is

$$X(x,z) = 1 + z \; b(x,z,x) \; . \qquad\qquad (9.53)$$

Similarly, the equation for Y(x,z) is

$$Y(x,z) = z \; h(x,z,x) \; . \qquad\qquad (9.54)$$

The equation for J is easily obtained by setting
v = -z in the decomposition formula, Eq. (9.52). The
result is

$$0 = - z^{-1} \, J(t,x,z) + \frac{\lambda}{4} \, X(x,z) \, b(t,-z,x)$$

$$- \frac{\lambda}{4} \, Y(x,z) \, h(t,-z,x) \; , \tag{9.55}$$

or

$$J(t,x,z) = \frac{\lambda}{4} \, z \, [X(x,z) \, b(t,-z,x)$$

$$- Y(x,z) \, h(t,-z,x)] \; . \tag{9.56}$$

The internal intensity function I can now be determined exclusively from b and h through the formula

$$(z^{-1} + v^{-1}) \, I(t,v,x,z)$$

$$= z \, v^{-1} \, \frac{\lambda}{4} \, [X(x,z) \, b(t,-z,x) - Y(x,z) \, h(t,-z,x)]$$

$$+ \frac{\lambda}{4} \, [1 + z \, b(x,z,x)] \, b(t,v,x) \tag{9.57}$$

$$- \frac{\lambda}{4} \, z \, h(x,z,x) \, h(t,v,x) \; .$$

Summarizing, we can say that the functions of interest in the monodirectional illumination problem, namely I, J, X and Y , are reducible to simple combinations of b and h functions for the isotropic illumination problem.

Let us illustrate this point by a numerical computation of X, Y, J and I for the case $\lambda = 1$, $x = 1$, $z = 0.5$, $v = 0.5$, and $t = 0.5$. From Table 9.4, we find the values

$$b(t,v,x) = 0.47436 ,$$

$$h(t,v,x) = 1.5721 ,$$

$$b(t,-z,x) = 1.5721 ,$$ (9.58)

$$h(t,-z,x) = 0.47436 .$$

From the formulas, we get

$$X(x,z) = 0.39351 ,$$

$$Y(x,z) = 0.12501 ,$$ (9.59)

$$J(t,x,z) = 0.27967 ,$$

and finally

$$I(t,v,x,z) = 0.13737 .$$ (9.60)

These values are all in complete agreement with values in-
dependently calculated.

Internal intensity functions for the case $\lambda = 1.0$,
$x = 2.0$ and $z = 0.5 = \cos 60^{\circ}$ are shown in Figure 9.13.
Each curve is for a fixed altitude, and the abscissa is the
angle of propagation, $\cos^{-1} v$ for $-1 < v \leq +1$.   Reflected
intensities are the intensities at the top in outgoing directions $(0 < v \leq 1)$. Some are shown in Figure 9.14 for the case
$\lambda = 0.5$, various thicknesses, and three different incident
directions.   These correspond to solutions of Riccati equations of the type discussed in Chapter 4.

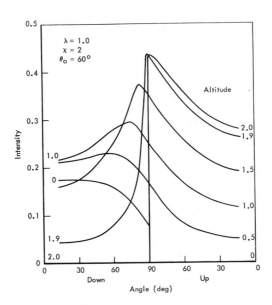

Figure 9.13   Some Internal Intensity Functions for
$\lambda = 1.0$, x = 2, $\Theta_o = 60°$

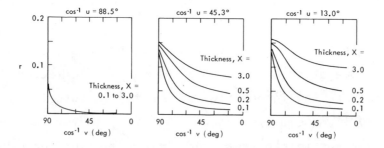

Figure 9.14   Reflected Intensities for $\lambda = 0.5$

9.8   X AND Y FUNCTIONS FOR LAMBERT'S LAW REFLECTOR

Another classical problem in multiple scattering of
radiation is that of determining the radiation field produced
by uniform parallel rays incident on a finite isotropically
scattering slab bounded by a Lambert's law reflector. It can
be treated by reduction to a Cauchy system involving an
auxiliary function of three variables (two angles and thickness). Here the problem is reduced to a Cauchy system involving auxiliary functions of merely one angular argument.
This is of analytic interest and computational utility. It
leads to the hope that similar advances may be possible in
the treatment of more realistic atmospheric models, and for
integral equations having other types of kernels.

Let uniform parallel rays of net flux π be incident
on a finite slab of optical thickness x . The incident rays
make an angle arccos z with respect to an inward drawn normal.
Within the slab scattering is isotropic and conservative, and
the slab is bounded on one side by a Lambert's law reflector
with albedo A . A Lambert's law reflector with albedo A
reflects the fraction A of radiation incident on it isotropically into all directions, for $0 \leq A \leq 1$. The source
function at optical altitude t (the rate of production of
scattered radiation per unit volume per unit solid angle
at optical altitude t) is denoted by J ,

$$J = J(t,x,u,A) , \tag{9.61}$$

$$0 \leq t \leq x, \quad 0 \leq u \leq 1, \quad 0 \leq A \leq 1 .$$

This function satisfies the Fredholm integral equation

$$J(t,x,u,A) = \frac{1}{4} e^{-(x-t)/u} + \frac{1}{2} A u e^{-x/u} E_2(t)$$

$$+ \int_0^x \left[\frac{1}{2} E_1(|t-y|) + AE_2(t) E_2(y) \right] \qquad (9.62)$$

$$\cdot \; J(y,x,u,A) \; dy \; ,$$

$$0 \leq t \leq x \; ,$$

where the functions E_1 and E_2 are defined by the integrals

$$E_1(r) = \int_0^1 e^{-r/z} \; dz/z \; , \qquad\qquad\qquad (9.63)$$

$$E_2(r) = \int_0^1 e^{-r/z} \; dz \; , \qquad\qquad r>0 \; . \quad (9.64)$$

Our goal is to reduce the determination of the function J to the solution of a Cauchy system involving J and several auxiliary functions of one angular argument. The independent variable of the Cauchy system will be x, the optical thickness of the medium.

DERIVATION. Differentiation of both sides of Eq. (9.62) with respect to x yields the relation

$$J_x(t,x,u,A) = - \frac{1}{u} \left[\frac{1}{4} e^{-(x-t)/u} + \frac{1}{2} A u e^{-x/u} E_2(t) \right]$$

$$+ \left[\frac{1}{2} E_1(x-t) + AE_2(t) E_2(x) \right] J(x,x,u,A) \quad (9.65)$$

$$+ \int_0^x \left[\frac{1}{2} E_1(|t-y|) + AE_2(t) E_2(y) \right] J_x(y,x,u,A) \; dy \; .$$

Regard this as an integral equation for the function J_x ,
the first two terms on the right hand side being the forcing
terms. Then keeping the definitions in Eqs. (9.63) and (9.64)
in mind, the solution of Eq. (9.65) is seen to be

$$J_x(t,x,u,A) = -\frac{1}{u} J(t,x,u,A)$$

(9.66)

$$+ 2J(x,x,u,A) \int_0^1 J(t,x,u',A) \, du'/u' ,$$

$$x \geq t .$$

This is one of the desired differential equations. Its use,
though, necessitates consideration of the auxiliary function
$J(x,x,u,A)$, the source function at $t = x$.

Introduce the optical depth τ into the integral
Eq. (9.62) by means of the relations

$$\tau = x-t ,$$

(9.67)

$$y' = x-y .$$

The result is

$$J(x-\tau,x,u,A) = \frac{1}{4} e^{-\tau/u} + \frac{1}{2} I(u,x,A) \, E_2(x-\tau)$$

$$+ \frac{1}{2} \int_0^1 E_1(|\tau-y'|) \, J(x-y',x,u,A) \, dy' ,$$

where I , the intensity of radiation leaving the Lambert's
law surface is

$$I(u,x,A) = A \left\{ ue^{-x/u} + 2 \int_0^x J(y,x,u,A) \, E_2(y) \, dy \right\}.$$
$$(9.69)$$

Differentiation of Eq. (9.68) with respect to x shows that

$$\frac{d}{dx} J(x-\tau,x,u,A) = \frac{1}{2} I_x(u,x,A) \, E_2(x-\tau)$$

$$+ \frac{1}{2} I(u,x,A) \frac{d}{dx} E_2(x-\tau)$$
$$(9.70)$$

$$+ \frac{1}{2} E_1(x-\tau) \, J(0,x,u,A)$$

$$+ \frac{1}{2} \int_0^x E_1(|\tau-y'\tau|) \frac{d}{dx} J(x-y',x,u,A) \, dy' \, .$$

Regarding this as an integral equation for the function $(d/dx) \, J(x-\tau,x,u,A)$ and recalling that

$$\frac{d}{dx} E_1(x) = -E_1(x) , \qquad\qquad x>0 , \quad (9.71)$$

we see that

$$\frac{d}{dx} J(x-\tau,x,u,A)$$

$$= \left[J(0,x,u,A) - I(u,x,A) \right] 2 \int_0^1 J(\tau,x,u') \, du'/u'$$
$$(9.72)$$

$$+ I_x(u,x,A) \, 2 \int_0^1 J(\tau,x,u') \, du' \, .$$

The function $J(t,x,u)$ is defined by the relation

$$J(t,x,u) = J(t,x,u,0) . \tag{9.73}$$

$$0 \le t \le x, \quad 0 \le u \le 1 .$$

and the function $J(t,x,u)$ itself satisfies the integral equation

$$J(t,x,u) = \frac{1}{4} e^{-(x-t)/u}$$

$$\tag{9.74}$$

$$+ \frac{1}{2} \int_0^x E_1 (|t-y|) \, J(y,x,u) \, dy ,$$

$$0 \le t \le x ,$$

which is a special case of Eq. (9.62) . Putting $\tau = 0$ yields the desired result,

$$\frac{d}{dx} J(x,x,u,A)$$

$$= \left[J(0,x,u,A) - I(u,x,A) \right] 2 \int_0^1 J(0,x,u') \, du'/u'$$

$$\tag{9.75}$$

$$+ I_x(u,x,A) \, 2 \int_0^1 J(0,x,u') \, du' .$$

Next we shall consider the function I defined in Eq. (9.69). Through differentiation we find that

$$I_x(u,x,A) = A \left\{ -e^{-x/u} + 2J(x,x,u,A) \, E_2(x) \right.$$

$$+ \; 2\int_0^1 \left\{ -\frac{1}{u} \, J(y,x,u,A) \right. \tag{9.76}$$

$$\left. + \; 2J(x,x,u,A)\int_0^1 J(y,x,u',A) \; du'/u' \right\} E_2(u) \; dy \right\}$$

or

$$I_x(u,x,A) = -\frac{1}{u} \, I(u,x,A)$$

$$+ \; 2J(x,x,u,A)\int_0^1 I(u',x,A) \; du'/u' \; . \tag{9.77}$$

This is the desired differential equation for the function I .

The differential equation for the function $J(0,x,u,A)$ follows from Eq. (9.66) by putting $t = 0$. It is

$$J_x(0,x,u,A) = -\frac{1}{u} \, J(0,x,u,A)$$

$$+ \; 2J(x,x,u,A)\int_0^1 J(0,x,u',A) \; du'/u' \; . \tag{9.78}$$

Lastly, we note that the functions $J(0,x,u)$ and $J(x,x,u)$ satisfy the differential equations

$$\frac{d}{dx} \, J(x,x,u) = J(0,x,u) \; 2\int_0^1 J(0,x,u') \; du'/u' \tag{9.79}$$

and

$$\frac{d}{dx} J(0,x,u) = -\frac{1}{u} J(0,x,u)$$

(9.80)

$$+ 2J(x,x,u)\int_0^1 J(0,x,u') \, du'/u' \, .$$

These equations follow from Eqs. (9.65) and (9.68) by putting $A = 0$ and from the fact that

$$I(u,x,0) = 0 \, ,$$

(9.81)

which is evident from Eq. (9.69).

The needed differential equations for the function $J(t,x,u,A)$ and the auxiliary functions $J(0,x,u,A)$, $J(x,x,u,A)$, $J(0,x,u)$, $J(x,x,u)$ and $I(u,x,A)$ have now been obtained. They are given in Eqs. (9.66), (9.78), (9.75), (9.80), (9.79) and (9.77). The initial conditions at $x = 0$ are

$$J(0,0,u,A) = \frac{1}{4} + \frac{1}{2} Au \, ,$$

(9.82)

$$J(0,0,u) = \frac{1}{4} \, ,$$

(9.83)

and

$$I(u,0,A) = Au \, .$$

(9.84)

These follow from Eqs. (9.62), (9.74) and (9.69).

STATEMENT OF THE CAUCHY SYSTEM. We introduce the standard
nomenclature

$$X(x,u) = 4J(x,x,u) \; , \tag{9.85}$$

$$Y(x,u) = 4J(0,x,u) \; , \tag{9.86}$$

as well as the new functions

$$X(x,u,A) = J(x,x,u,A) \; , \tag{9.87}$$

$$Y(x,u,A) = J(0,x,u,A) \; . \tag{9.88}$$

The Cauchy system for these four functions and the fifth
function I is

$$X_x(x,u) = \frac{1}{2} \; Y(x,u) \int_0^1 Y(x,u') \; du'/u' \; , \tag{9.89}$$

$$Y_x(x,u) = - \frac{1}{u} Y(x,u) + \frac{1}{2} X(x,u) \int_0^1 Y(x,u') \; du'/u' \; , \tag{9.90}$$

$$X_x(x,u,A) = \frac{1}{2} \left[Y(x,u,A) -1(u,x,A) \right] \int_0^1 Y(x,u') \; du'/u'$$

$$+ \frac{1}{2} \left\{ - \frac{1}{u} I(u,x,A) + 2X(x,u,A) \right. \tag{9.91}$$

$$\left. \cdot \int_0^1 I(u',x,A) \; du'/u' \right\} \int_0^1 Y(x,u') du'/u',$$

$$Y_x(x,u,A) = -\frac{1}{u} Y(x,u,A)$$

$$(9.92)$$

$$+ 2X(x,u,A) \int_0^1 Y(x,u',A)\ du'/u' \ ,$$

and

$$I_x(u,x,A) = -\frac{1}{u} I(u,x,A)$$

$$(9.93)$$

$$+ 2X(x,u,A) \int_0^1 I(u',x,A)\ du'/u' \ ,$$

$$x \geq 0;$$

$$X(0,u) = 1 \ , \qquad\qquad\qquad\qquad (9.94)$$

$$Y(0,u) = 1 \ , \qquad\qquad\qquad\qquad (9.95)$$

$$X(0,u,A) = \frac{1}{4} + \frac{1}{2} Au \ , \qquad\qquad (9.96)$$

$$Y(0,u,A) = \frac{1}{4} + \frac{1}{2} Au \ , \qquad\qquad (9.97)$$

$$(u,0,A) = Au \ . \qquad\qquad\qquad (9.98)$$

The Cauchy system for the function $J(t,x,u,A)$, valid for $x \geq t$, is

$$J_x(t,x,u,A) = -\frac{1}{u} J(t,x,u,A)$$

$$(9.99)$$

$$+ 2X(x,u,A) \int_0^1 J(t,x,u',A)\ du'/u' \ ,$$

$$x \geq t \ ,$$

the initial condition at x = t being

$$J(t,t,u,A) = X(t,u,A) \ .$$ (9.100)

DISCUSSION. The Cauchy system presented in the previous
section lends itself well to numerical solution via the
method of lines. The integrals are approximated via finite
sums using Gaussian quadrature formulas. Thus the differential-integral equations are reduced to ordinary differential equations with known initial conditions. Results
of successful calculations for the functions X(x,u) and
Y(x,u) have already been presented. Calculations of the
source function J(t,x,u,A) can readily be made. Results
for source functions for the case in which the reflector is
a specular reflector are available following the theory
developed in Chapter 5 on composite kernels.

Several tasks remain. It is to be proved that the
solution of the Cauchy system which we have derived does
provide a solution of the integral equation. Finally, we
note that it would be desirable to express the scattering
function S and the transmittion function T in terms of
the auxiliary functions discussed in this paper.

9.9   INVERSE PROBLEMS IN RADIATIVE TRANSFER

Problems of radiative transfer in planetary and
stellar atmospheres have been extensively treated by many
investigators. Much of this work has dealt with what we may
call *direct problems*, i.e., the determination of the

intensities of radiation produced by certain light incident
on a medium with known scattering and absorbing properties.
Recently, though, there has been increasing interest in
inverse problems. In these, the aim is to deduce the nature
of the medium (and often the source) on the basis of measurements of the radiation field produced by incident
ratiation.

We shall study inverse problems in radiative transfer
via the method of quasilinearization. We wish to show that
such problems can be viewed as multipoint boundary-value
problems for large systems of nonlinear ordinary differential equations. These equations can then be resolved numerically using current digital computers and the method of
quasilinearization.

We consider a slab which consists of two layers of
isotropically scattering material of unknown optical thickness and albedo for single scattering. Given the measured
intensity of the light diffusely reflected from the slab, we
wish to determine the optical thickness of the two layers
and their albedos for single scattering.

We consider an inhomogeneous, plane-parallel, non-
emitting, and isotropically scattering atmosphere of finite
optical thickness τ_1 whose optical properties depend only
upon τ, the optical height above the bottom $(0 \leq \tau \leq \tau_1)$.
Let parallel rays of light of net flux π per unit area
normal to their direction of propagation be incident on the
upper surface in the direction characterized by the number
μ_0 $(0 < \mu_0 < 1)$, where μ_0 is the cosine of the angle measured from the normal to the surface. The bottom surface
is a completely absorbing barrier, so that no light is reflected from it.

The intensity of the diffusely reflected light in the direction $\cos^{-1} \mu$ is $r(\mu,\mu_0,\tau_1)$. We define a related function $R(\mu,\mu_0,\tau_1)$, symmetric in μ and μ_0, by writing

$$r(\mu,\mu_0,\tau_1) = R(\mu,\mu_0,\tau_1)/4\mu . \tag{9.101}$$

The function R satisfies the integro-differential equation

$$\frac{\partial R(\mu,\mu_0,\tau_1)}{\partial \tau_1} = -\left(\frac{1}{\mu} + \frac{1}{\mu_0}\right) R(\mu,\mu_0,\tau_1)$$

$$+ \lambda(\tau_1) \left[1 + \frac{1}{2} \int_0^1 R(\mu,\mu',\tau_1) \frac{d\mu'}{\mu'}\right] \tag{9.102}$$

$$\cdot \left[1 + \frac{1}{2} \int_0^1 R(\mu'',\mu_0,\tau_1) \frac{d\mu''}{\mu''}\right],$$

where $\lambda(\tau_1)$ is the albedo for single scattering, and R is subject to the condition

$$R(\mu,\mu_0,0) = 0 . \tag{9.103}$$

This equation is obtained by means of the theory of invariant imbedding. A discrete version of Eqs. (9.102) and (9.103) is obtained by replacing the ingegrals by finite sums using a Gaussian quadrature formula of order N. Let us introduce new functions $R_{ij}(\tau_1)$,

$$R_{ij}(\tau_1) = R(\mu_i,\mu_j,\tau_1); \tag{9.104}$$

$$i,j=1,2,\ldots,N,$$

where $\{\mu_i\}$ is the set of N roots of the shifted Legendre polynomial $P_N^*(\mu) = P_N(1-2\mu)$. The values of μ_i and the corresponding Christoffel weights have been tabulated for $N = 3, 4, \ldots, 15$. The functions $R_{ij}(\tau_1)$ satisfy the system of ordinary nonlinear differential equations

$$\frac{dR_{ij}(\tau_1)}{d\tau_1} = -\left(\frac{1}{\mu_i} + \frac{1}{\mu_j}\right) R_{ij}(\tau_1)$$

$$+ \lambda(\tau_1)\left[1 + \frac{1}{2}\sum_{k=1}^{N} R_{ik}(\tau_1)\frac{w_k}{\mu_k}\right], \quad (9.105)$$

$$\cdot \left[1 + \frac{1}{2}\sum_{k=1}^{N} R_{kj}(\tau_1)\frac{w_k}{\mu_k}\right],$$

with the initial conditions

$$R_{ij}(0) = 0, \quad\quad\quad\quad\quad\quad (9.106)$$
$$\text{for } i=1,2,\ldots,N; j=1,2,\ldots,N.$$

AN INVERSE PROBLEM. Let us now consider a medium composed of two layers, with albedo for single scattering λ_1 in the lower layer and albedo λ_2 in the upper layer $(\lambda_1 \neq \lambda_2)$. The total optical thickness of this slab is T and the thickness of the lower layer is c. We wish to determine c, T, and the parameters λ_1 and λ_2.

Let us assume that we have obtained N^2 (noisy) measurements of the diffusely reflected light, b_{ij}, where b_{ij} is the intensity of the light diffusely reflected in the direction μ_i caused by incident parallel rays of net flux π in the direction μ_j. The constants which characterize

this medium, λ_1, λ_2, c, and T are to be determined so
that the theoretical diffuse reflection pattern produced
by using the estimated values in the differential Eqs. (9.105)
will agree as closely as possible, in the least-squares sense,
with the observed field $\{b_{ij}\}$; i.e., we wish to minimize the
expression

$$\sum_{i=1}^{N} \sum_{j=1}^{N} [r_{ij}(T) - b_{ij}]^2 . \qquad (9.107)$$

Let us assume that the albedo function has the form

$$\lambda(\tau) = a + b \tanh 10(\tau - c) \qquad (9.108)$$

so that

$$\lambda_1 \cong a - b \text{ in Layer 1,}$$
$$\qquad\qquad\qquad\qquad\qquad\qquad\qquad (9.109)$$
$$\lambda_2 \cong a + b \text{ in Layer 2,}$$

and c is the position of the boundary between Layer 1
and Layer 2.

The "observations" $\{b_{ij}\}$ are produced computationally with the use of Eq. (9.108) for the albedo function,
where we set

$$a = 0.5, \quad b = 0.1, \quad c = 0.5 , \qquad (9.110)$$

and with the use of the differential Eqs. (9.105), integrating out to a thickness

$$T = 1.0 .$$

The observations which this produces are given in Table 9.7.

NUMERICAL SOLUTION VIA QUASILINEARIZATION.   Let us consider
the problem posed above in more general and flexible terms.
An R-dimensional column vector $x(t)$ is a solution of the
differential equation

$$\dot{x} = f(x), \tag{9.111}$$
$$0 \leq t \leq T.$$

The value of T is assumed known. This again represents
no essential restriction, as we shall see. We wish to determine the initial vector c,

$$x(0) = c , \tag{9.112}$$

in such a manner that we minimize the quadratic expression

$$(x(T) - b, x(T) - b) = Q , \tag{9.113}$$

where $x(T)$ is the solution of Eq. (9.111) for t = T. The
first s components of c are required to be zero, and the
remaining ones are variables to be determined.

Our computational formalism proceeds as follows:
First, an initial approximation to the desired initial vector in selected. We then proceed inductively. Suppose that
we have obtained a kth approximation to c, which we denote
by c^k, and a kth approximation to the solution function
$x^k(t)$. Note that the superscripts refer to the order of
the approximation and not to the components of c and x.

Table 9.7

The Measurements $\{b_{ij}\}$

	j=1	2	3	4	5	6	7
i = 1	0.079914	0.028164	0.014304	0.009104	0.006707	0.005515	0.004970
2	0.143038	0.091522	0.058437	0.040826	0.031405	0.026378	0.023989
3	0.167000	0.134331	0.099653	0.075106	0.060044	0.051445	0.047248
4	0.178898	0.157955	0.126408	0.099392	0.081253	0.070435	0.065042
5	0.185284	0.170817	0.142072	0.114229	0.094495	0.082423	0.076332
6	0.188723	0.177733	0.150791	0.122665	0.102104	0.089349	0.082870
7	0.190354	0.180898	0.154995	0.126773	0.105829	0.092748	0.086083

We obtain the vectors c^{k+1} and $x^{k+1}(t)$ in this manner. The vector $x^{k+1}(t)$ is a solution of the *linear* system in x^{k+1}

$$\dot{x}^{k+1} = f(x^k) + J(x^k)(x^{k+1} - x^k) , \qquad (9.114)$$

where $J(x^k)$ is the Jacobian matrix with elements

$$J_{ij} = (\partial f_i / \partial x_j) . \qquad (9.115)$$

We produce a particular solution $p(t)$ of the system (9.114) by numerically integrating the system

$$\dot{p} = f(x^k) + J(x^k)(p - x^k) , \qquad (9.116)$$

where $p(t)$ is subject to the initial condition

$$p(0) = 0 . \qquad (9.117)$$

Then we produce numerically $R - s$ independent solutions of the homogeneous system

$$\dot{h}_i = J(x^k) h_i , \qquad (9.118)$$

where the vector $h_i(t)$ is subject to the initial condition

$$[j^{th} \text{ component of } h_i(0)] = \delta_{ij} , \qquad (9.119)$$

$$i = s+1, \ldots, R,$$

δ_{ij} being the Kronecker delta function. These integrations are readily carried out on the interval $0 \leq t \leq T$, since complete sets of initial conditions are given. From general theory we know that the vector x^{k+1} is representable in the form

$$x^{k+1}(t) = p(t) + \sum_{i=s+1}^{R} m_i h_i(t) , \qquad (9.120)$$

where the numbers m_i are arbitrary constants. In view of the way we have chosen the initial conditions in Eqs. (9.117) and (9.119), we see that m_i represents the initial value of the ith component of the vector x^{k+1}. In particular, we shall have, in numerical form, the values of the vectors $p(T)$, $h_{s+1}(T),\ldots,h_R(T)$.

We now wish to minimize the expression

$$\left(p(T) + \sum_{i=s+1}^{R} m_i h_i(T) - b , \right.$$

$$\left. p(T) + \sum_{i=s+1}^{R} m_i h_i(T) - b \right)$$

$$= F(m_{s+1}, m_{s+2}, \ldots, m_R) \qquad (9.121)$$

over the values of $m_i, i = s + 1, s + 2, \ldots, R$, where b is known experimentally and $p(T)$, and $h_{s+1}(T), \ldots, h_R(T)$ are known computationally. This minimization is done by solving, numerically, the linear algebraic system of equations .

$$\partial F/\partial m_i = 0, \tag{9.122}$$

$$i = s+1, s+2, \ldots, R.$$

The values of m_{s+1}, m_{s+2}, $\ldots$, m_R so obtained are the values of the last $R - s$ components of the vector c^{k+1}, and $x^{k+1}(t)$ can be determined from Eq. (9.120) using the values of m_{s+1}, $m_{s+2} \ldots, m_R$ and the stored values of $p(t)$ and $h_i(t)$, $i = s + 1$, $s + 2$, $\ldots, R$.

The entire process is then repeated to obtain the $(k + 2)$nd approximation.

To simplify the procedure of finding the m's, let us observe that the relation

$$x^{k+1}(T) \cong b \tag{9.123}$$

can be written in the form

$$\sum_{i=s+1}^{R} h_i(T) m_i \cong b - p(T) . \tag{9.124}$$

If we let H be the matrix whose i^{th} column is $h_i(T)$, $i = s + 1$, $s + 2$, $\ldots$, R, y be the column vector whose i^{th} component is m_i, then the above relation can be rewritten

$$Hm \cong y , \tag{9.125}$$

where H is a matrix with R rows and $R - s$ columns, and m and y are $R - s$ dimensional column vectors. According to the theory of the method of least squares the vector m which minimizes the expression

$$(Hm - y, Hm - y) = G \qquad\qquad (9.126)$$

is

$$m = (H'H)^{-1} H'y , \qquad\qquad (9.127)$$

where H' is the transpose of the matrix H.

A few general comments on the procedure are now in order. The selection of a good initial approximation is important, since in this case the method is rapidly convergent, with the number of correct digits approximately doubling with each additional step; if, however, the initial approximation is too poor, the method may be divergent. At each step we must integrate R(R - s + 1) first order differential equations with given initial conditions to produce the R - s homogeneous solutions and the one particular solution. In addition we must solve a system of R - s linear algebraic equations. This may be a sizable computing load, and the solving of the linear algebraic equations can be a source of great difficulty.

NUMERICAL EXPERIMENTS TO DETERMINE THE THICKNESS OF THE LOWER LAYER. In the first series of experiments we wish to determine only the thickness of the lower layer, assuming that all of the other parameters of the medium are known. We use a seven-point Gaussian quadrature, so that N = 7. We consider c to be a function of τ_1 for which $dc/d\tau_1 = 0$. Following the method prescribed in the preceding section, we obtain $N^2 = 49$ linear differential equations for R_{ij}^{k+1} and one equation for c^{k+1}. Thus there are 50 linear equations

$$
\frac{dR_{ij}^{k+1}}{d\tau_1} = \Bigg\{ - \left(\frac{1}{\mu_i} + \frac{1}{\mu_j}\right) R_{ij}^{k}
$$

$$
+ \lambda(\tau_1) \left(1 + \frac{1}{2} \sum_{\ell=1}^{N} R_{i\ell}^{k} \frac{w_\ell}{\mu_\ell}\right)
$$

$$
\cdot \left(1 + \frac{1}{2} \sum_{\ell=1}^{N} R_{\ell j}^{k} \frac{w_\ell}{\mu_\ell}\right)\Bigg\}
$$

$$
+ \Bigg\{ - \left(\frac{1}{\mu_i} + \frac{1}{\mu_j}\right) (R_{ij}^{k+1} - R_{ij}^{k})
$$

$$
+ \frac{1}{2} \lambda(\tau_1) \Bigg[\left(1 + \frac{1}{2} \sum_{\ell=1}^{N} R_{i\ell}^{k} \frac{w_\ell}{\mu_\ell}\right)
$$

$$
\cdot \sum_{\ell=1}^{N} (R_{\ell j}^{k+1} - R_{\ell j}^{k}) \frac{w_\ell}{\mu_\ell}
$$

$$
+ \left(1 + \frac{1}{2} \sum_{\ell=1}^{N} R_{\ell j}^{k} \frac{w_\ell}{\mu_\ell}\right)
$$

$$
\cdot \sum_{\ell=1}^{N} (R_{i\ell}^{k+1} - R_{i\ell}^{k}) \frac{w_\ell}{\mu_\ell} \Bigg]\Bigg\}
$$

$$
+ \Bigg\{ (c^{k+1} - c^{k}) \left(1 + \frac{1}{2} \sum_{\ell=1}^{N} R_{i\ell}^{k} \frac{w_\ell}{\mu_\ell}\right)
$$

$$
\cdot \left(1 + \frac{1}{2} \sum_{\ell=1}^{N} R_{\ell j}^{k} \frac{w_\ell}{\mu_\ell}\right)
$$

$$
\cdot [- 10b\ \mathrm{sech}^2\ 10(\tau_1 - c^{k})]\Bigg\}\ , \qquad (9.128)
$$

$$
i,j = 1,2,\ldots,N,
$$

and

$$\frac{dc^{k+1}}{d\tau_1} = 0 \; , \tag{9.128'}$$

where

$$\lambda(\tau_1) = a + b \; \tanh \; 10(\tau_1 - c^k) \; .$$

One may reduce the number of differential equations in the system (9.128) by using the symmetry property

$$R_{ij}^{\;k+1}(\tau_1) = R_{ji}^{\;k+1}(\tau_1) \; , \tag{9.129}$$

$$i=1,2,\ldots,N, \quad j=1,2,\ldots,N.$$

Thus instead of $N^2 + 1 = 50$ differential equations for $R_{ij}^{\;k+1}$ and c^{k+1}, we have $[N(N + 1)/2] + 1 = 29$ differential equations plus the finite Eqs. (9.129) to fill in the missing values of R. However, we will still speak of the matrix $\{R_{ij}^{\;k+1}\}$ as having 7 rows and 7 columns.

Now let the 50-dimensional vextor $x^{k+1}(\tau_1)$ represent the 49 elements $R_{ij}^{\;k+1}(\tau_1)$ taken in some order,

$$x_\ell^{\;k+1}(\tau_1) = R_{ij}^{\;k+1}(\tau_1) \tag{9.130}$$

for $\ell=1,2,\ldots,49$ as $i=1,2,\ldots,7$, and $j=1,2,\ldots,7$, and

$$x_{50}^{\;k+1}(\tau_1) = c^{k+1}(\tau_1) \; . \tag{9.131}$$

We express $x^{k+1}(\tau_1)$ as a sum of a particular solution $p(t)$ and a homogeneous solution $h(t)$ of the Eqs. (9.128),

$$x^{k+1}(\tau_1) = p(\tau_1) + mh(\tau_1) . \tag{9.132}$$

We require that the multiplier m be chosen to minimize the expression

$$\sum_{\ell=1}^{49} [p_\ell(1) + mh_\ell(1) - b_\ell]^2 , \tag{9.133}$$

where the singly subscripted set $\{b_\ell\}$ ($\ell=1,2,\ldots,49$) represents the doubly subscripted set of constants $\{4\mu_i b_{ij}\}$ ($i=1,2,\ldots,7$; $j=1,2,\ldots,7$) taken in the same order as the transformation (9.130). The value of m which minimizes (9.133) is

$$m = \sum_{\ell=1}^{49} h_\ell(1)[b_\ell - p_\ell(1)]/\sum_{\ell=1}^{49} [h_\ell(1)]^2 , \tag{9.134}$$

as we see from a simple differentiation. With the proper choice of initial values $p(0)$ and $h(0)$, this constant m is the initial value

$$m = c^{k+1}(0)$$

and hence approximates directly the thickness of Layer 1. The results of three experiments with different initial guesses of the thickness c are shown in Table 9.8.

Table 9.8

Successive Approximations of c

Approximation	Run 1	Run 2	Run 3
0	0.2	0.8	0.0
1	0.62	0.57	No
2	0.5187	0.5024	convergence
3	0.500089	0.499970	
4	0.499990	0.499991	
True Value 0.5		0.5	0.5

The initial approximation is generated in each run by integrating the nonlinear Eqs. (9.105) with the value of c listed as approximation zero.

Even with 60 percent errors in the initial approximations, the value of c is determined to one part in fifty thousand or 0.002 percent in four iterations. The time required on the IBM 7044 is 1-1/2 min/run, using an integration step size of 0.01 and a fourth order Adams-Moulton method. Making use of the symmetry of R, 58 linear differential equations have to be integrated for the particular and homogeneous solutions for each approximation. In Run 3 the solution diverges because the initial guess is not good enough.

NUMERICAL EXPERIMENTS TO DETERMINE T, THE OPTICAL THICKNESS. Next we consider the estimation of the total optical thickness T when all of the other system constans are known and

49 measurements {b} are given. The formulation of the linear boundary-value problem proceeds as for the previous case with one major difference. The terminal boundary for the interval of integration is unknown. In order that the interval be fixed, we define a new independent variable σ,

$$\sigma T = \tau_1 \,,$$

with $0 \leq \sigma \leq 1$ as the range of integration. The constant T is a solution of the equation

$$dT/d\sigma = 0 \,. \tag{9.135}$$

Then one may obtain the linear equations which correspond to (9.128), with independent variable σ instead of τ_1.
    Three trials are made to determine T, with initial guesses $T^0 = 0.9$, 1.5, and 0.5 respectively. Recall that the true value is T = 1.0. Within 4 min of computing time, four iterations are carried out per trial, and the results obtained are correct to one part in 100,000.

NUMERICAL EXPERIMENTS TO DETERMINE λ_1, λ_2, AND c, THE ALBEDOS AND OPTICAL THICKNESS. In the final experiment, on the basis of the 49 observations, we try to determine λ_1, λ_2, and c assuming T = 1.0 is known. Again we use N = 7. This time there are three homogeneous solutions and a particular solution to compute, each with $N(N + 1)2 + 3 = 31$ components, so that $4 \times 31 = 124$ linear differential equations are to be integrated during each stage of the successive approximation calculations. A standard (Gaussian elimination) matrix inversion procedure is used to invert

the 3 x 3 matrix of the linear algebraic system. The results in Table 9.9 are obtained in three iterations which
consume 2 min of computing time on the IBM 7044 machine.

Table 9.9

Successive Approximations of λ_1, λ_2, and c

Approximation	λ_1	λ_2	c
0	0.51	0.69	0.4
1	0.4200	0.6052	0.5038
2	0.399929	0.599995	0.499602
3	0.399938	0.599994	0.499878
True Values	0.4	0.6	0.5

DISCUSSION. The general conceptual and computational approach to inverse problems which we have discussed here is
by no means limited to the simple physical model considered.
At the expense of additional computing time we may treat
atmospheres having anisotropic scattering, time-dependent
sources, and so on. We may also wish to consider wave
rather than particle processes. Applications to orbit determination, cardiology, and other areas have been made.
Of particular importance is the question of the effect of
errors in the observations on the accuracy of the estimates
of the parameters. In this connection it appears that the
use of the min-max criterion rather than that of least
squares may be efficacious.

9.10   DISCUSSION

This chapter has shown a direct application of imbedding concepts to the theory of radiative transfer, where
integral equations abound. While Cauchy systems may be derived analytically beginning with the integral equations, we
have chosen to present here the physical type of derivation,
at least in the earlier sections of this chapter. It would
be interesting to see physical derivations made in other
applications as well. We have seen how well the imbedding
method works, and have numerous graphs and tables to prove
it.

The b and h functions introduced in a general
setting in Chapter 4 are identified as internal intensity
functions for the case of isotropic illumination of a slab.
In terms of these functions, all of the relevant functions
of the monodirectional slab problem can be expressed using
simple algebraic formulas.

For the slab problem with a Lambert's law reflector,
the kernel has a certain composite form. The point of interest
here is that the auxiliary function of three variables can be
replaced by several functions of fewer variables, thus reducing the computing load. It also leads us to attempt to
extend the reduction to other types of kernels.

Another class of basic problems in radiative transfer
is that of inverse problems. An inverse problem might require the estimation of the structure of a scattering medium
based on measurements of reflected radiation. This is the
subject of the preceding section.

REFERENCES

1. V. V. Sobolev, *A Treatise on Radiative Transfer*, D. Van Nostrand Co., Princeton, New Jersey, 1963.

2. R. E. Bellman, H. H. Kagiwada, R. E. Kalaba and S. Ueno, "Numerical Results for Chandrasekhar's X and Y Functions of Radiative Transfer", *Journal of Quantitative Spectroscopy and Radiative Transfer*, v. 6, (1966), pp. 479-500.

3. H. H. Kagiwada, *System Identification: Methods and Applications*, Addison-Wesley, Reading, Mass., 1974.

4. R. W. Preisendorfer, *Radiative Transfer on Discrete Spaces*, Pergamon Press, New York, 1965.

Chapter 10

DUAL INTEGRAL EQUATIONS

10.1   INTRODUCTION

A system of differential equations with initial values will be presented, and it will be shown that a solution of the Cauchy system is also a solution of a system of dual integral equations.

The dual integral equations are

$$\int_0^\infty A(\lambda) \; J_0(\lambda r) \; d\lambda = f(r), \qquad\qquad r<a, \quad (10.1)$$

$$\int_0^\infty \frac{\lambda A(\lambda)}{1-g(\lambda)} \; J_0(\lambda r) \; d\lambda = 0, \qquad\qquad r>a, \quad (10.2)$$

where J_0 is the usual Bessel function, the functions f and g are given, and the function A is to be determined. Such equations occur in potential theory and elsewhere.

10.2   VALIDATION OF THE INITIAL VALUE PROBLEM

In this section, it will be proved that under approp-
riate hypotheses the function $e(a,z)$, $a \geq 0$, $z \geq 0$ defined by
the system of differential equations with initial values

$$R_a(a,z,z') = \left[\cos za + \frac{2}{\pi} \int_0^\infty g(u) \cos ua \, R(a,z,u) \, du \right]$$

$$\cdot \left[\cos z'a + \frac{2}{\pi} \int_0^\infty g(u) \cos ua \, R(a,u,z') \, du \right], \qquad a \geq 0, \quad (10.3)$$

$$R(0,z,z') = 0, \qquad\qquad\qquad\qquad z,z' \geq 0, \quad (10.4)$$

$$e_a(a,z) = \left[F(a) + \frac{2}{\pi} \int_0^\infty g(u) \cos ua \, e(a,u) \, du \right]$$

$$\cdot \left[\cos za + \frac{2}{\pi} \int_0^\infty g(u) \cos ua \, R(a,z,u) \, du \right], \qquad a \geq 0, \quad (10.5)$$

$$e(0,z) = 0, \qquad\qquad\qquad\qquad z \geq 0, \quad (10.6)$$

satisfies the system of dual integral equations

$$\int_0^\infty (1-g(z)) \, e(a,z) \, J_0(zr) \, dz = \int_0^{\pi/2} F(r \sin \theta) \, d\theta, \qquad r < a, \quad (10.7)$$

$$\int_0^\infty ze(a,z) \, J_0(zr) \, dz = 0, \qquad\qquad r > a. \quad (10.8)$$

The subscript   a   in Equations (10.3) and (10.5) denotes differentiation.   This convention will be followed through this
section.

It will be assumed that the functions involved in the
subsequent discussions are such that the various changes of
the orders of integration and differentiation performed are
valid.

We shall now establish three lemmas.

Lemma I

Under the hypothesis that the differential system in
ρ

$$\rho_a(a,z,z') = \left[\cos z'a + \frac{2}{\pi} \int_0^\infty g(u) \cos au\, R(a,u,z')\, du \right]$$

$$\cdot \left[\cos za + \frac{2}{\pi} \int_0^\infty g(u) \cos au\, \rho(a,z,u)\, du \right],$$
$$a \geq 0, \quad (10.9)$$

$$\rho(0,z,z') = 0, \qquad\qquad\qquad z,z' \geq 0, \quad (10.10)$$

has a unique solution, the function $R(a,z,z')$, solution of
the differential system (10.3) with initial values (10.4),
satisfies the integral equation

$$R(a,z,z') = \frac{\pi}{2} K(a,z,z') + \int_0^\infty K(a,z,u)\, R(a,u,z')\, g(u)\, du,$$
$$(10.11)$$

where

$$K(a,z,z') = \frac{2}{\pi} \int_0^a \cos tz \cos tz' \, dt \; . \qquad (10.12)$$

Proof

Let $\rho(a,z,z')$ be defined by

$$\rho(a,z,z') = \frac{\pi}{2} K(a,z,z') + \int_0^\infty K(a,z,u) \, R(a,u,z')g(u) \, du. \qquad (10.13)$$

Differentiation of Equation (10.13) and use of Equation (10.12) yields

$$\rho_a(a,z,z')$$

$$= \frac{\pi}{2} K_a(a,z,z') + \int_0^\infty K_a(a,z,u) \, R(a,u,z')g(u) \, du$$

$$+ \int_0^\infty K(a,z,u) \, R_a(a,u,z') \, g(u) \, du = \cos za \cos z'a$$

$$+ \frac{2}{\pi} \int_0^\infty \cos za \cos ua \, R(a,u,z') \, g(u) \, du$$

$$+ \int_0^\infty K(a,z,u) \, R_a(a,u,z') \, g(u) \, du$$

$$= \cos za \left[\cos z'a + \frac{2}{\pi} \int_0^\infty g(u) \cos ua \, R(a,u,z') \, du \right]$$

$$+ \int_0^\infty K(a,z,u) \, R_a(a,u,z') \, g(u) \, du \; . \qquad (10.14)$$

The last term on the right-hand side can be further analyzed
using Equation (10.3):

$$\int_0^\infty K(a,z,u) \; R_a(a,u,z') \; g(u) \; du$$

$$= \int_0^\infty K(a,z,u) \left[\cos au + \frac{2}{\pi} \int_0^\infty g(v) \cos va \right.$$
$$\left. \cdot \; R(a,u,v) \; dv \right]$$

$$\cdot \left[\cos az' + \frac{2}{\pi} \int_0^\infty g(v) \cos va \; R(a,v,z') \; dv \right] g(u) \; du$$

$$= \left[\cos az' + \frac{2}{\pi} \int_0^\infty g(v) \cos va \; R(a,v,z') \; dv \right]$$

$$\cdot \int_0^\infty K(a,z,u) \left[\cos au + \frac{2}{\pi} \int_0^\infty g(v) \cos va \; R(a,u,v) \; dv \right]$$
$$\cdot \; g(u) \; du \; . \tag{10.15}$$

Furthermore, using the definition of ρ in Equation (10.13),

$$\int_0^\infty K(a,z,u) \left[\cos au + \frac{2}{\pi} \int_0^\infty g(v) \cos va \, R(a,u,v) \, dv \right] g(u) \, du$$

$$= \int_0^\infty K(a,z,u) \cos au \, g(u) \, du$$

$$+ \frac{2}{\pi} \int_0^\infty g(u) \, K(a,z,u) \int_0^\infty g(v) \cos va \, R(a,u,v) \, dv \, du$$

$$= \int_0^\infty K(a,z,u) \cos au \, g(u) \, du$$

$$+ \frac{2}{\pi} \int_0^\infty g(v) \cos va \int_0^\infty g(u) \, K(a,z,u) \, R(a,u,v) \, du \, dv$$

$$= \int_0^\infty K(a,z,v) \cos av \, g(v) \, dv$$

$$+ \frac{2}{\pi} \int_0^\infty g(v) \cos va \int_0^\infty g(u) \, K(a,z,u) \, R(a,u,v) \, du \, dv$$

$$= \frac{2}{\pi} \int_0^\infty g(v) \cos va \left[\frac{\pi}{2} K(a,z,v) \right.$$

$$\left. + \int_0^\infty g(u) \, K(a,z,u) \, R(a,u,v) \, du \right] dv$$

$$= \frac{2}{\pi} \int_0^\infty g(v) \cos va \, \rho(a,z,v) \, dv \, . \qquad\qquad (10.16)$$

Substitution in Equation (10.14) using the results of Equations (10.15) and (10.16) shows that the function $\rho(a,z,z')$ satisfies the differential relation (10.9). Since clearly

$$K(0,z,z') = 0 , \tag{10.17}$$

we have

$$\rho(0,z,z') = 0 . \tag{10.18}$$

Since the system given by Equations (10.9) and (10.10) are obviously satisfied by $R(a,z,z')$, the uniqueness hypothesis provides

$$\rho(a,z,z') = R(a,z,z') . \tag{10.19}$$

This completes the proof of Lemma 1.

Lemma 2

The function e satisfies the Fredholm integral equation

$$e(a,z) = \int_0^a F(t) \cos zt \, dt$$

$$+ \int_0^\infty K(a,z,u) \, g(u) \, e(a,u) \, du , \tag{10.20}$$

where the kernel $K(a,z,u)$ is defined by Equation (10.12).

Proof

Let the function $\psi(a,z)$ be defined by the relation

$$\Psi(a,z) = \int_0^a F(t) \cos zt \, dt$$

$$+ \int_0^\infty K(a,z,u) \, g(u) \, e(a,u) \, du \qquad (10.21)$$

$$a \geq 0, \ z \geq 0.$$

Differentiation of Equation (10.21) and use of Equation (10.12) yields

$$\Psi_a(a,z) = F(a) \cos za + \int_0^\infty K_a(a,z,u) \, g(u) \, e(a,u) \, du$$

$$+ \int_0^\infty K(a,z,u) \, g(u) \, e_a(a,u) \, du$$

$$= \cos za \left[F(a) + \frac{2}{\pi} \int_0^\infty g(u) \cos ua \, e(a,u) \, du \right]$$

$$+ \int_0^\infty K(a,z,u) \, g(u) \, e_a(a,u) \, du \, . \qquad (10.22)$$

The last term can be further analyzed by use of Equations (10.5) and (10.11) (*Lemma 1*) :

$$\int_0^\infty K(a,z,u) \; g(u) \; e_a \, (a,u) \; du = \int_0^\infty K(a,z,u) \; g(u)$$

$$\cdot \left[F(a) + \frac{2}{\pi} \int_0^\infty g(v) \; \cos \, va \; e(a,v) \; dv \right]$$

$$\cdot \left[\cos \, au + \frac{2}{\pi} \int_0^\infty g(v) \; \cos \, va \; R(a,u,v) \; dv \right] du$$

$$= \left[F(a) + \frac{2}{\pi} \int_0^\infty g(v) \; \cos \, va \; e(a,v) \; dv \right]$$

$$\cdot \int_0^\infty K(a,z,u) \left[\cos \, au + \frac{2}{\pi} \int_0^\infty g(v) \right.$$

$$\cdot \; \cos \, va \; R(a,u,v) \; dv \bigg] \; g(u) \; du$$

$$= \left[F(a) + \frac{2}{\pi} \int_0^\infty g(v) \; \cos \, va \; e(a,v) \; dv \right]$$

$$\cdot \left[\int_0^\infty K(a,z,v) \; g(v) \; \cos \, av \; dv \right.$$

$$+ \frac{2}{\pi} \int_0^\infty g(v) \; \cos \, av \int_0^\infty g(u) \; K(a,z,u)$$

$$\cdot \; R(a,u,v) \; du \; dv \bigg] = \left[F(a) + \frac{2}{\pi} \int_0^\infty g(v) \; \cos \, va \; e(a,v) \, dv \right]$$

(Over)

$$\cdot \left[\frac{2}{\pi} \int_0^\infty g(v) \ \cos \ va \left(\frac{\pi}{2} \ K(a,z,v) \right. \right.$$

$$\left. \left. + \int_0^\infty g(u) \ K(a,z,u) \ R(a,u,v) \ du \right) dv \right]$$

$$= \left[F(a) + \frac{2}{\pi} \int_0^\infty g(v) \ \cos \ va \ e(a,v) \ dv \right]$$

$$\cdot \left[\frac{2}{\pi} \int_0^\infty g(v) \ \cos \ av \ R(a,z,v) \ dv \right] . \qquad (10.23)$$

Equations (10.22) and (10.23) show that

$$e(a,z) = \psi_a(a,z) , \qquad\qquad (10.24)$$

Furthermore

$$\psi(0,z) = 0 , \qquad\qquad (10.25)$$

since

$$K(0,z,z') = 0 . \qquad\qquad (10.26)$$

This completes the proof of Lemma 2.

Lemma 3

Let $\phi(t)$ be differentiable in the interval $[0,a]$ of the real line. Then

$$\int_0^\infty z J_0(zr) \int_0^a \phi(t) \ \cos \ zt \ dt \ dz = 0, \quad r>a. \qquad (10.27)$$

Integration by parts shows that

$$\int_0^a \phi(t) \cos zt \, dt = \phi(a) \frac{\sin za}{z} - \int_0^a \phi'(t) \frac{\sin zt}{z} \, dt.$$

(10.28)

Using this we may write

$$\int_0^\infty zJ_0(zr) \int_0^\infty \phi(t) \cos zt \, dt \, dz$$

$$= \int_0^\infty \phi(a) \, J_0(zr) \sin za \, dz$$

$$- \int_0^a \phi'(t) \int_0^\infty J_0(zr) \sin zt \, dz \, dt \, .$$

(10.29)

The thesis follows from the relation

$$\int_0^\infty J_0(zr) \sin zt \, dz = 0 \, , \qquad\qquad r>t \, . \quad (10.30)$$

The main theorem may be now stated and proved.

Theorem

Under the assumptions of Lemmas 1 and 2, the function e(a,z) that satisfies Equations (10.5) and (10.6) satisfies also the dual integral Equations (10.7) and (10.8),

Proof of Equation (8).

The following derivation uses the results of Lemmas 2 and 3:

$$\int_0^\infty ze(a,z) \, J_0(zr) \, dz = \int_0^\infty z\left(\int_0^a F(t) \, \cos zt \, dt\right.$$

$$+ \left.\int_0^\infty K(a,u,z) \, g(u) \, e(a,u) \, du\right)J_0(zr) \, dz$$

$$= \int_0^\infty zJ_0(zr) \int_0^a F(t) \, \cos zt \, dt \, dz$$

$$+ \int_0^\infty zJ_0(zr) \int_0^\infty K(a,u,z) \, g(u) \, e(a,u) \, du \, dz,$$
$$r>a. \quad (10.31)$$

The right hand side can be shown to equal zero by use of Lemma 3:

$$\int_0^\infty zJ_0(zr) \int_0^\infty K(a,z,u) \, g(u) \, e(a,u) \, du \, dz$$

$$= \int_0^\infty zJ_0(zr)\left(\int_0^\infty \frac{2}{\pi}\int_0^\infty \cos tu \, \cos tz \, dt \, g(u) \, e(a,u) \, du\right)d$$

$$= \frac{2}{\pi}\int_0^\infty zJ_0(zr)\int_0^\infty \cos tz \left(\int_0^\infty g(u) \, e(a,u) \, \cos tu \, du\right)dtd$$

$$= 0 \, . \quad\quad\quad\quad\quad\quad\quad\quad\quad (10.32)$$

Proof of Equation (10.7).

Let

$$I_1 = \int_0^\infty e(a,z) \, J_0(zr) \, dz \, ,$$

(10.33)

and

$$I_2 = \int_0^\infty g(z) \, e(a,z) \, J_0(zr) \, dz;$$

(10.34)

then

$$\int_0^\infty (1-g(z)) \, e(a,z) \, J_0(zr \, dz = I_1 - I_2.$$

(10.35)

Using Lemma 2 to study I_1 it is seen that

$$I_1 = \int_0^\infty e(a,z) \, J_0(zr) \, dz$$

$$= \int_0^\infty \left(\int_0^a F(t) \cos zt \, dz + \int_0^\infty K(a,z,u) \, g(u) \, e(a,u) \, du \right)$$

$$\cdot \, J_0(zr) \, dz = \int_0^a F(t) \int_0^\infty J_0(zr) \cos zt \, dz \, dt$$

$$+ \int_0^\infty \int_0^\infty K(a,z,u) \, g(u) \, e(a,u) \, du \, J_0(zr) \, dz$$

(10.36)

$$= \int_0^\infty F(t) \, (r^2 - t^2)^{-1/2} \, dt$$

$$+ \int_0^\infty \int_0^\infty K(a,z,u) \, g(u) \, e(a,u) \, du \, J_0(zr) \, dz,$$

where the relation

$$\int_0^\infty J_0(zr) \cos zt \, dz = \begin{cases} 0, & t>r \\ (r^2-t^2)^{-1/2}, & r>t \end{cases}, \qquad (10.37)$$

has been used for the first term in the right hand side. Substituting $t = r \sin \phi$ in that term we find

$$\int_0^r F(t)(r^2-t^2)^{-1/2} \, dt = \int_0^{\pi/2} F(r \sin \phi) \, d\phi. \qquad (10.38)$$

To analyze the second term in the right hand side of Equation (10.36) we use the relation

$$J_0(zr) = \frac{2}{\pi} \int_0^{\pi/2} \cos (zr \sin \phi) \, d\phi. \qquad (10.39)$$

Using the definition of $K(a,z,u)$ given in equation (10.12) that term becomes

$$\int_0^\infty \int_0^\infty K(a,z,u) \; g(u) \; e(a,u) \; du \; J_0(zr) \; dz$$

$$= \frac{2}{\pi} \int_0^\infty \int_0^\infty \left(\int_0^a \cos tz \cos tu \; dt \right) g(u) \; e(a,u) \; du J_0(zr) dz$$

$$= \frac{2}{\pi} \int_0^\infty g(u) \; e(a,u) \left(\int_0^a \cos tu \left(\int_0^\infty J_0(zr) \cos zt \; dz \right) dt \right) du$$

$$= \frac{2}{\pi} \int_0^\infty g(u) \; e(a,u) \left(\int_0^\infty \cos tu \; (r^2-t^2)^{-1/2} dt \right) du$$

$$= \int_0^\infty g(u) \; e(a,u) \left(\frac{2}{\pi} \int_0^{\pi/2} \cos(ur \sin\phi) d\phi \right) du$$

$$= \int_0^\infty g(u) \; e(a,u) \; J_0(ur) du = I_2 \; . \qquad\qquad (10.40)$$

Equation (10.36) together with Equations (10.38) and (10.40) shows that

$$I_1 = \int_0^{\pi/2} F(r \sin \phi) \; d\phi + I_2 \; . \qquad\qquad (10.41)$$

The result follows from Equation (10.35) and (10.41).

EXERCISE

1. Derive the Cauchy system from the dual integral equations.

REFERENCES

1. J. Buell, H. Kagiwada, R. Kalaba, E. Ruspini, and
    E. Zagustin, "Solution of a System of Dual Integral
    Equations," *International J. Engineering Science*,
    Vol. 10 (1972), pp. 503-510.
2. R. Kalaba and E. Zagustin, "An Initial Value Method for
    Dual Integral Equations," *International J. Engineering
    Science*, Vol. 10 (1972), pp. 603-608.

INDEX